VOLUME FIFTY TWO

Advances in
PHYSICAL ORGANIC
CHEMISTRY

ADVISORY BOARD

VOLUME FIFTY TWO

Advances in
PHYSICAL ORGANIC
CHEMISTRY

Edited by

IAN H. WILLIAMS
Department of Chemistry,
University of Bath,
Bath, United Kingdom

NICHOLAS H. WILLIAMS
Department of Chemistry,
University of Sheffield,
Sheffield, United Kingdom

ELSEVIER

ACADEMIC PRESS
An imprint of Elsevier

Academic Press is an imprint of Elsevier
125 London Wall, London, EC2Y 5AS, United Kingdom
The Boulevard, Langford Lane, Kidlington, Oxford OX5 1GB, United Kingdom
525 B Street, Suite 1650, San Diego, CA 92101, United States
50 Hampshire Street, 5th Floor, Cambridge, MA 02139, United States

First edition 2018

Notices
Knowledge and best practice in this field are constantly changing. As new research and
experience broaden our understanding, changes in research methods, professional practices,
or medical treatment may become necessary.

Practitioners and researchers must always rely on their own experience and knowledge in
evaluating and using any information, methods, compounds, or experiments described
herein. In using such information or methods they should be mindful of their own safety and
the safety of others, including parties for whom they have a professional responsibility.

To the fullest extent of the law, neither the Publisher nor the authors, contributors, or editors,
assume any liability for any injury and/or damage to persons or property as a matter of
products liability, negligence or otherwise, or from any use or operation of any methods,
products, instructions, or ideas contained in the material herein.

ISBN: 978-0-12-815211-9
ISSN: 0065-3160

For information on all Academic Press publications
visit our website at https://www.elsevier.com/books-and-journals

Working together
to grow libraries in
developing countries

www.elsevier.com • www.bookaid.org

Publisher: Zoe Kruze
Acquisition Editor: Jason Mitchell
Editorial Project Manager: Peter Llewellyn
Production Project Manager: Vignesh Tamil
Cover Designer: Victoria Pearson

Typeset by SPi Global, India

CONTENTS

CONTRIBUTORS

Luke Anderson
Department of Chemistry, University of Liverpool, Liverpool, United Kingdom

Roman Boulatov
Department of Chemistry, University of Liverpool, Liverpool, United Kingdom

Jason B. Harper
School of Chemistry, University of New South Wales, Sydney, NSW, Australia

Rebecca R. Hawker
School of Chemistry, University of New South Wales, Sydney, NSW, Australia

Hans-Ullrich Siehl
Institute of Organic Chemistry I, Universität Ulm, Ulm, Germany

PREFACE

The aim of *Advances in Physical Organic Chemistry* is to provide the chemical community with authoritative and critical assessments of the many aspects of this broad area of chemical science. To define physical organic chemistry in terms of the scope of its content is a difficult task, perhaps because it is better considered as a way of thinking about and actually doing chemistry. The scope of the present volume 52 embraces carbocations, ionic liquids, and polymers within its diversity, but the chapters are unified by means of their common application of the techniques and methodology of physical organic chemistry to the study of disparate topics.

Hans–Ullrich Siehl gives us a coherent account of experimental and computational investigations on the structure and dynamics of bicyclobutonium $C_4H_7^+$ and related carbocations showing us how, in particular, calculations of NMR spectroscopic properties have contributed to solve this long-standing structural conundrum and have helped to establish the bonding principles of hypercoordinated carbon in many electron-deficient compounds.

Rebecca Hawker and Jason Harper discuss recent advances in understanding of solvent effects in ionic liquids and how these determine rates and selectivities of reactions carried out in these increasing popular media. They show how knowledge of the properties of ionic liquids may be utilized predictively in preparative chemistry, and point out the opportunities that this new understanding may present into the future.

In their review of polymer mechanochemistry, Luke Anderson and Roman Boulatov describe what happens when macromolecular chains are subjected to mechanical loads and how to understand this in molecular terms. Whereas much effort has been focused on creating polymers which undergo complex and interesting reactions, including mechanochromism and load strengthening, less progress has been achieved in creating a theoretically sound framework for the organization, systematization, and generalization of existing manifestations of polymer mechanochemistry, and to guide the design of new mechanochemical systems. They provide a critical and constructive analysis of the area, highlighting the impact and opportunities of the physical organic approach to provide a deeper understanding of this field.

May this volume help us not only to see a little further but to open our eyes to new perspectives, with the help of our contributing authors.

CHAPTER ONE

The Conundrum of the $(C_4H_7)^+$ Cation: Bicyclobutonium and Related Carbocations

Hans-Ullrich Siehl[1]

Institute of Organic Chemistry I, Universität Ulm, Ulm, Germany
[1]Corresponding author: e-mail address: ullrich.siehl@uni-ulm.de

Dedication: This chapter is dedicated to George A. Olah (1927–2017), the sole Nobel Laureate in Chemistry of 1994, thanking him for 45 years of continuous encouragement and friendship.[1]

Contents

Advances in Physical Organic Chemistry, Volume 52
ISSN 0065-3160
https://doi.org/10.1016/bs.apoc.2018.10.001

Abstract

Selected retrospective and recent investigations on the structure and dynamics of bicyclobutonium cations $(C_4H_7)^+$ and related carbocations are reviewed. Practically all the tools, methods, and techniques of experimental physical organic chemistry have been used to investigate these cation systems. Bench-type chemistry such as solvolysis experiments, kinetic measurements, product analysis, and measurement of kinetic isotope effects and also IR spectroscopy in cryogenic matrices and gas-phase studies have been applied. Multinuclear NMR spectroscopy in super acidic solution introduced by G.A. Olah and CP-MAS solid-state NMR studies at cryogenic temperatures have been important experimental contributions. The investigation of equilibrium isotope effects on NMR spectra of fast-equilibrating systems introduced by Saunders has revealed decisive details on structure and dynamics of bicyclobutonium ions. High level high quantum-chemical calculations of structure, energies and, in particular, calculation of NMR spectroscopic properties such as chemical shifts and nuclear spin–spin coupling constants have contributed to solve the conundrum of bicyclobutonium ions. All the results on bicyclobutonium ions finally helped to firmly establish the bonding principle of hypercoordinated carbon in these and many other electron-deficient compounds.

Prologue

"Among nonclassical ions the ratio of conceptual difficulty to molecular weight reaches a maximum with the cyclopropyl–cyclobutyl cation $(C_4H_7)^+$ system."[2]

1. INTRODUCTION

The cyclobutyl/cyclopropylmethyl cation system $(C_4H_7)^+$ has most likely been the focus of more studies than any other carbocation system except the 2-norbornyl cation.[3–5] This contribution reviews selected retrospective and more recent results on $(C_4H_7)^+$ cations such as 1–bicyclo[1.1.0] butonium ions and related cations.

Different methods and technologies have been applied over a timespan of more than half a century to enlighten the structure and dynamics of this type of cations. Practically all the tools of experimental physical organic chemistry have been utilized to shed light on the conundrum of the $(C_4H_7)^+$ cation system. The methods and techniques range from typical bench–type wet chemistry, such as kinetic measurements and product analysis of solvolysis reactions, including the measurement of kinetic isotope effects (KIEs), various techniques of gas–phase chemistry, IR spectroscopy, multinuclear NMR spectroscopy in super acidic solution and CP-MAS solid–state NMR, investigation of equilibrium isotope effects (EIEs) on NMR spectra of fast-equilibrating systems, and last but not least high level quantum-chemical calculations of structure, energies and, in particular, NMR spectroscopic properties such as chemical shifts and nuclear spin–spin coupling constants.

This contribution is not intended to be a complete survey of the topic. The selection is made primarily to illustrate the different aspects of the subject. For the line of argument, it has been considered to discuss some recent tools and methods more thoroughly than others.

1.1 Early Studies

Cyclobutyl substrates **1** and cyclopropylmethyl substrates **2** solvolyse at high rates and give similar substitution products **3**, **4**, and **5** (Fig. 1).[6] The observed mixture of cyclobutyl- (**3**), cyclopropylmethyl- (**4**), and minor amounts of homoallyl- (**5**) substitution products could result from a single cationic intermediate $(C_4H_7)^+$ **6** that may be attacked at different sites by a nucleophile or may involve two or more rapidly equilibrating cations. The structure and dynamics of the intermediate cation $(C_4H_7)^+$ has been an intriguing problem for more than half a century.

Several isomeric structures have been proposed to account for the experimental results. The structures considered are the bisected cyclopropylmethyl cation **7**, a puckered cyclobutyl cation **8**, a bridged hypercoordinated bicyclobutonium ion **9**, a tricyclobutonium ion **10**, and the homoallyl cation **11** (Fig. 2). The puckered cyclobutyl cation **8** and the bicyclobutonium ion **9** are related in symmetry and distinguished only according to whether the transannular carbons are in bonding distance or not.

The bicyclobutonium cation **9** has a pentacoordinated C_γ carbon. It can be called a protonated bicyclobutane. The bridging interaction between C_α and C_γ can formally be drawn as an interaction of the backside of the C_γ—H_{endo} orbital with the formally empty carbon p-orbital at C_α as shown in **12**.[7]

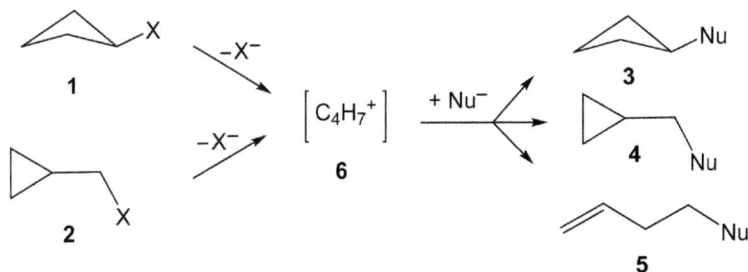

Fig. 1 Solvolysis of cyclobutyl and cyclopropylmethyl substrates.

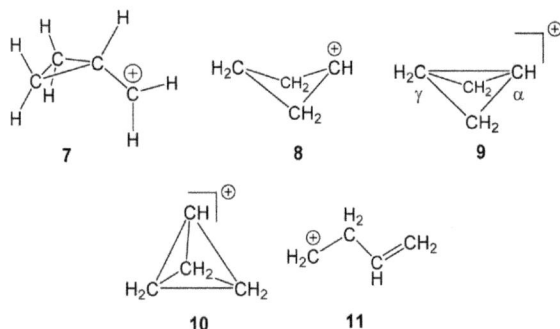

Fig. 2 Isomeric $(C_4H_7)^+$ structures.

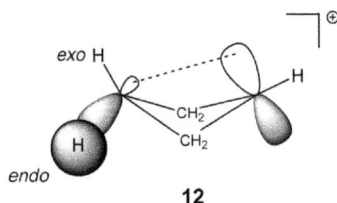

All experimental and computational evidence indicates that the open homoallyl cation **11** is significantly higher in energy than the other structures and does not contribute significantly to the structures involved in the equilibrium of the parent $(C_4H_7)^+$ cation system. The threefold symmetrical structure, the tricyclobutonium **10**, at first suggested to explain the experimental results,[8] was later ruled out. Product distribution studies of S_N1 reactions of specifically ^{14}C-labeled cyclobutyl- and cyclopropylmethyl-precursor compounds **1** and **2** did not show full equilibration of the tag as expected from a threefold symmetric intermediate tricyclobutonium ion **10**. Therefore a set of rapidly equilibrating threefold–degenerate bicyclobutonium ions was suggested.[9] In any case **10** would be higher in energy and would be expected to distort because of the Jahn Teller effect.[10]

1.2 Terminology

The coining of the term "*nonclassical*" ionic intermediate goes back to the communication of Roberts and Mazur in 1951 where it was first used to describe the tricyclobutonium cation structure.[8] In his 1990 autobiography, Roberts gives some historical details on the term nonclassical:

> The structure **10** (and also **9**) was so unprecedented and bizarre that I christened it, then and there a "nonclassical" carbocation. This designation caught on like

gangbusters, for no obvious reason. Indeed it became a generic description for unusual bridged organic intermediates … It was not a very precise term, and perhaps that was part of its appeal.[11]

Paul Bartlett at first considered the name "nonclassical ion" as infelicitous and later in his review book *Nonclassical Ions* he wrote: "After protesting for years against the inappropriate name '*nonclassical Ions*' I have been over-ruled by general usage and am employing the term because of its of its extreme familiarity."[2]

The term was later used by Winstein to describe the structure of the bridged 2-norbornyl cation.[12] It was this cation which caused a historical "classical—nonclassical" ion controversy.[13]

Beside the dispute over the correct structure of the 2-norbornyl cation the term "nonclassical ion" initiated quite a lot of confusion. It caused some damage as it led to partial loss of reputation of physical organic chemistry. A recent X-ray crystallographic structure determination of the 2-norbornyl cation can be regarded as the final close of the norbornyl cation ion controversy.[14]

The term "*nonclassical ion*" initiated new forms of drawing chemical structures on paper. Sometimes ambiguous and obscure "dotted line" representations of chemical structures caused a dispute over representational issues with what we mean when we write chemical structures on paper. The successful convention of drawing chemical structures in Lewis-type formalism, where a single bond is made up of two electrons and is represented as a solid line, is no longer applicable to "non-Lewis-type" chemical structures. "It has been recognized for almost a century that conventional bond formulas with single, double, and triple bonds are not adequate to account for the geometries and reactions of many substances."[15,16]

There are at least three books on "*nonclassical organic structures.*"[2,4,17] Although the term "nonclassical" entity is not precise it is still being used today.[18,19] The author of this review prefers to avoid the term because of its ambiguity. Avoiding dotted lines, preference is given here to write solid lines to represent bonds even if they have lower occupancy than two electrons.

An electron-deficient bonding principle is better described as hypercoordinative bonding. A carbon atom is hypercoordinated if the number of attached atoms exceeds four. A carbon can be hypercoordinated, i.e., have more than the usual four covalent bonds, without violation of the octet-rule when the bonds on average involve fewer than two electrons.[20] The same applies to hydrogen-hypercoordination, i.e., more than one covalent bond to a hydrogen, where the bonds involve less than two electrons.

Hypercoordination and hyperconjugation are related phenomena. In the parlance of MO-theory, both bonding schemes are described by multicenter two-electron bonds. These are most commonly three-center two-electron (3c–2e) bonds. However, hyperconjugation does not generally lead to such strong changes of geometry as are observed for bridged hypercoordinated molecules. There is no clear boundary between the two effects; they are considered as regions in the spectrum of structural variation because of charge delocalization in carbocations.

1.3 Kinetic Isotope Effects

A major difficulty in understanding the ionic intermediates involved in solvolysis and other S_N1-type reactions is that the nature of the intermediate cations involved and their interconversion is not directly accessible but is inferred from rate studies and product analyses. Ion-pair phenomena during nucleophilic substitution reactions and solvolysis of cyclopropylmethyl and cyclobutyl derivatives may reflect on the product-formation step. Ion pairs do not behave like free carbocations in their reactions with nucleophiles. Therefore, any conclusions about the structure and the nature of the cations which are based upon the composition of solvolysis products may not be unambiguous.[21]

Deuterium KIEs have always been a powerful tool to investigate reaction mechanisms and have allowed some light to be shed on the structure of transition states.[22] Investigation of secondary deuterium KIEs effects in the solvolysis of cyclobutyl methanesulfonates showed a reduced α-effect, an inverse β-effect, and a rather large normal γ-effect. This indicated anchimeric assistance by strong 1–3 interaction in the transition state in the rate-determining step, in accord with a 1,3-bridged bicyclobutonium like structure of the transition state.[23]

1.4 Gas-Phase Studies

Gas-phase reactivity studies of $(C_4H_7)^+$ cations by collisional activated dissociation mass spectroscopy suggest that the bicyclobutonium ion **9** and the cyclopropylmethyl cation **7**, which are initially generated from corresponding precursors such as **1** and **2**, share a common reactivity outlet.[24,25] The experimental observations are in agreement with the predicted reactivity of the bicyclobutonium ion **9** as originally postulated by Roberts and coworkers.[9] Utilizing FT-ICR mass spectrometry and high-pressure radiolytic techniques, the equilibration of $(C_4H_7)^+$ cations has been studied: consistent with condensed-phase results and theoretical estimates, this study points to a remarkably close stability of isomeric ions and to their very fast $(10^{-10}$ s) equilibration time, suggesting a very low barrier to isomerization.[26]

1.5 Vibrational Spectroscopy

Cryogenic matrix IR spectroscopy of $(C_4H_7)^+$ cations generated from cyclopropylmethyl or cyclobutyl precursors,[27] indicates the presence of two isomeric ions of nearly equal energy. The most probable structures were proposed to be the bicyclobutonium ion **9** and the delocalized bisected cyclopropylmethyl cation **7**.

1.6 Structures and Equilibrium of Bicyclobutonium and Cyclopropylmethyl Cations

Different dotted line formulations including symmetric and nonsymmetric structures for bicyclobutonium ions such as **13** and **14** have been suggested.[9,15]

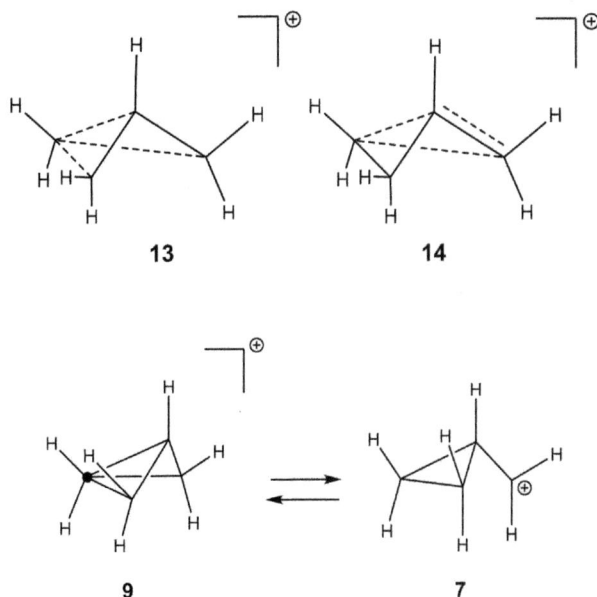

Fig. 3 Interconversion of isomeric cations **9** and **7**.

Even after decades of detailed study, the $(C_4H_7)^+$ cation system remained somewhat enigmatic. The current consensus[26] seems to be that the symmetrical bicyclobutonium cation **9** (point group C_s) and the bisected cyclopropylmethyl cation **7** exist as rapidly interconverting structures of nearly equal stability and that a very flat potential energy surface (PES) leads to a fast interconversion of these cations over a very low barrier (Fig. 3).

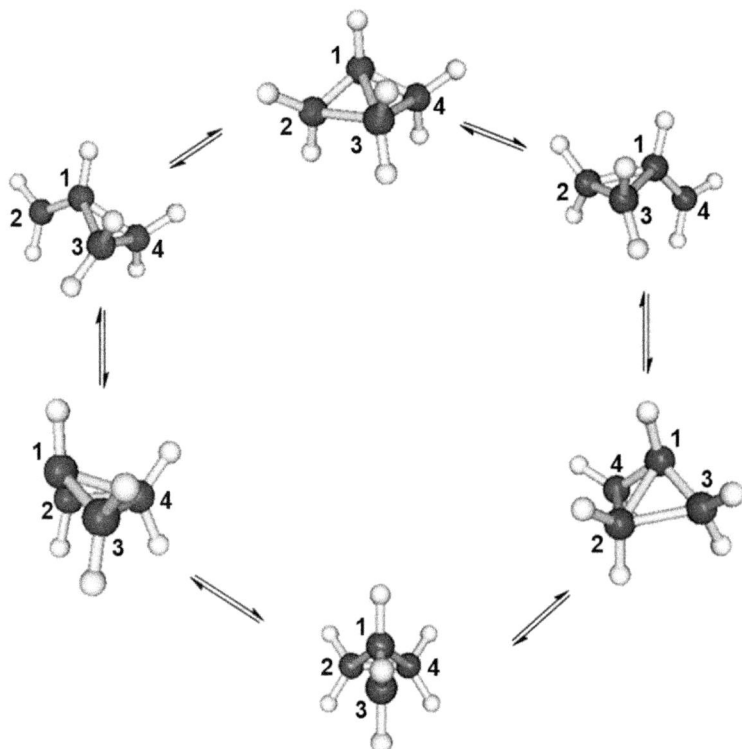

Fig. 4 Threefold-degenerate methylene rearrangement of bicyclobutonium ions **9** and cyclopropylmethyl cations **7**.

In a threefold-degenerate methylene rearrangement, three bicyclo-butonium ion structures and three cyclopropylmethyl structures can inter-convert (Fig. 4).

For both the parent $(C_4H_7)^+$ system and for the 1–methyl-substituted $[C_4H_6CH_3]^+$ cations, it has been ambiguous (and a controversially discussed question) as to whether (a) the bicyclobutonium and the cyclopropylmethyl structures are major or minor isomers of stable cations with slightly different stability or (b) one of them is a transition structure (TS) for the threefold-degenerate rearrangement of the other.[4]

1.7 NMR Spectroscopic Investigations

Experimental NMR spectroscopic investigations of $(C_4H_7)^+$ cations generated in solution as long-lived intermediates in superacidic media from cyclobutyl substrates and cyclopropylmethyl substrates led to the same

NMR spectra and afforded additional insight into structure and dynamics of the $(C_4H_7)^+$ cation system.[7]

Assignment of the experimental ^1H and ^{13}C NMR spectra to a single structure is not straightforward because the $(C_4H_7)^+$ cations are highly fluxional molecules undergoing fast rearrangements on a very flat energy surface. Each of the two structures to consider, the bicyclobutonium ion **9** and the cyclopropylmethyl cation **7** have two kinds of methylene carbon, but only a single ^{13}C NMR peak at ~ 52 ppm is seen for the methylene groups because of rapid averaging. No line broadening could be observed at the lowest accessible temperature in solution. The signal for the CH–carbon appears at ~ 112 ppm. The ^1H NMR shows that the six hydrogens at the three methylene groups are averaged as two separate sets of three hydrogens at 4.64 and 4.21 ppm, keeping the geminal-related *exo*- and *endo*-hydrogens distinct. This indicates that the rearrangement process which averages the methylene groups in the $(C_4H_7)^+$ cation is stereospecific. The nonequivalence of the geminal hydrogens shows that no fast conformational cyclobutyl ring inversion for the parent $(C_4H_7)^+$ cation is taking place. A cyclobutyl cation structure without significant C_α–C_γ bridging was excluded as a contributor to the equilibrium of $(C_4H_7)^+$ cations.[4,28]

The analysis of the temperature dependence of the ^{13}C NMR signals of the $(C_4H_7)^+$ cation in solution gave evidence for the presence of two, one major and one minor, isomeric $(C_4H_7)^+$ cations. At lower temperatures, the methine carbon signal is deshielded whereas the peak of the averaged CH_2-carbons is shielded. By comparison with calculated ^{13}C NMR chemical shifts, it was suggested that the bicyclobutonium ion **9** is a major isomer and the bisected cyclopropylmethyl cation **7** is a minor isomer in the fast dynamic equilibrium of $(C_4H_7)^+$ cations.[28]

Solid-state CP-MAS ^{13}C NMR at cryogenic temperatures gave additional strong support for two $(C_4H_7)^+$ isomers with nearly identical energy and a very flat energy surface.[29] The interpretation of the experiment is somewhat complicated by lattice effects in the solid state. Below 60 K, four distinct signals at ~ 235, 55, ~ 28, and ~ -15 ppm were observed. The intensity of the signal at 235 ppm, attributed to the formally positively charged methylene carbon of the cyclopropylmethyl cation **7**, decreases and the strongly shielded peak at -15 ppm, attributed to the pentacoordinated methylene carbon of the bicyclobutonium ion **9**, increases when the temperature is lowered to 5 K. The chemical shifts experimentally observed in the solid state for the formally positively charged methylene carbon of the cyclopropylmethyl cation at 235 ppm and for the shielded peak

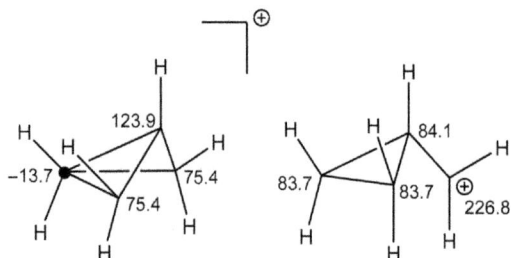

Fig. 5 CCSD/cc-pVTZ//CCSD(T)/qz2p calculated ^{13}C NMR chemical shifts for bicyclobutonium ion **9** and cyclopropylmethyl cation **7**.

at -15 ppm, attributed to the pentacoordinated methylene carbon of the bicyclobutonium ion are in reasonable accord with the corresponding calculated ^{13}C NMR chemical shifts recently recalculated at coupled cluster level for the structure (CCSD/cc-pVTZ) and at CCSD(T)/qz2p level for pristine isolated static structures of the bicyclobutonium **9** and the cyclopropylmethyl cation **7** (Fig. 5).[30]

1.8 EIEs on NMR Spectra of Fast-Rearranging (C$_4$H$_7$)$^+$ Cations

EIEs are also important experimental tools for the investigation of molecular processes.[22] The investigation of deuterium EIEs on NMR spectra of fast-equilibrating systems, after some sparsely earlier reports,[31] was mainly developed by Saunders and called "isotopic perturbation method of equilibria."[32,33] It was referred to as one of the most elegant experiments of classical physical organic chemistry.[34] The method has been used extensively as a sensitive probe of structure and dynamics of fast degenerate rearrangement equilibria of carbocations under stable-ion conditions in solutions, as well as for many other questions in organic and inorganic chemistry, to distinguish between a symmetrical structure and multiple energy minima of degenerate fast-equilibrating structures of lower symmetry.[35]

Isotope effects on a chemical equilibrium are observed when there is force constant change between the initial and the final state for a coordinate involving the isotopic atom. EIEs in NMR spectra of fast degenerate equilibrating systems ($K=1$), which have been desymmetrized by isotopic substitution by deuterium, proves a multiminimum PES, leading to time-averaged symmetry because of averaged signals of exchanging sites under conditions of fast exchange. In addition, EIE splittings in NMR spectra provide some general features of the exchange characteristics of the system under study. This can be evaluated from the NMR spectra of the partially deuterated systems preferentially admixed with some nondeuterated compound. The method is independent of whether the slow exchange region for the dynamic process is accessible or not. The lifting of the degeneracy of the equilibrium ($K_{HD} \neq 1$) leads to strongly temperature-dependent isotopic splitting patterns of the averaged signals in the NMR spectra. These EIE splitting patterns, corrected for temperature-independent and generally small intrinsic isotope shifts on NMR chemical shifts, allow the determination of the number of exchanging sites and their relative population. In most cases the direction of an EIE ($K_{HD} > 1$ or $K_{HD} < 1$) can be extracted from the NMR spectra. The temperature-dependent equilibrium constant K and the thermodynamic parameters ΔH^0 and ΔS^0 can be extracted from variable-temperature NMR spectra. The direction of the isotope effect gives important information on the relative stiffness of vibrational force constants of C—H bonds, which in turn can be related to characteristic features of the alternative structures under consideration.

The ^1H and ^{13}C NMR measurements of the isotopic perturbation of the fast equilibrium of deuterated $(C_4H_7)^+$ cations perturbed by one, two, or three deuterium atoms are likely to be the most decisive piece of experimental evidence for the preference of the hypercoordinated bridged structure of the bicyclobutonium **9** in solutions of super acids.[36–39] The investigation of triply deuterated ions[36] allowed a specific *endo-* and *exo*-assignment of the deuterium substitution. The ^1H and ^{13}C NMR spectra of the monodeuterated $(C_4H_6D)^+$ cation give clear evidence that the threefold-degenerate fast equilibrium of $(C_4H_7)^+$ cations takes place between a singly populated site which is shielded in a static ion and a doubly populated site which is deshielded in a static cation relative to the averaged methylene signal in the d_0 cation. Some evidence for cyclopropylmethyl cation **7** as a minor contributor to the $(C_4H_7)^+$ cation equilibrium has also been found. The nonbridged cyclobutyl structure **8** as well as the tricyclobutonium structure **10** was shown not to contribute to the equilibration process.

In the ^{13}C NMR spectra of a mixture of $(C_4H_7)^+$ and methylene-dideuterated $(C_4H_5D_2)^+$ cations, the signal of the two nondeuterated methylene groups in $(C_4H_5D_2)^+$ exhibits temperature-dependent shielding compared to the peak of the d_0 cation, indicating a definite EIE. Owing to low signal-to-noise ratio and longer relaxation time, the deshielded CD_2 signal is not visible (Fig. 6A). In the ^1H NMR of a mixture of $(C_4H_5D_2)^+/(C_4H_7)^+$ cations only the *exo*–CH_2 peak of the CD_2-labeled cation shows a temperature shielding whereas the *endo*–CH_2-signal shows no significant effect (Fig. 6B).

A 1:1 mixture of *exo*– and *endo*-methylene-monodeuterated $(C_4H_7D)^+$ cations generated from α-monodeuterated cyclopropyl methanol gave more detailed insight into the bonding situation of the main isomer of the $(C_4H_7)^+$ cation system. The two *exo*– and *endo*-$(C_4H_6D)^+$ isomers show temperature-dependent EIE splittings for the signal of the averaged methylene carbons which are different in sign and very different in magnitude. The splitting patterns in the ^1H and ^{13}C NMR spectra (Fig. 7) of the signals of the

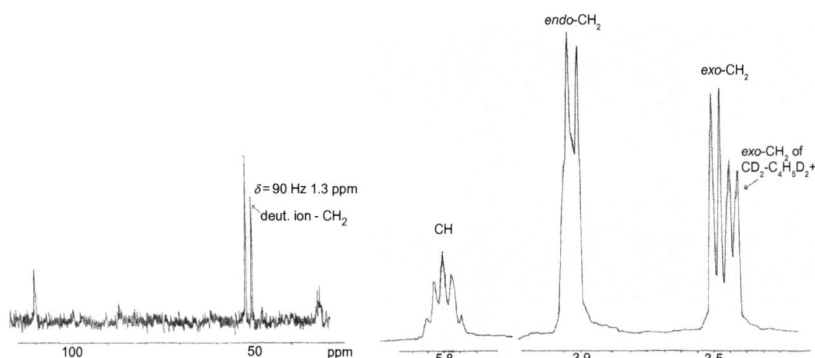

Fig. 6 (A) ^{13}C NMR and (B) ^1H NMR spectra of a $(C_4H_5D_2)^+/(C_4H_7)^+$ cation mixture.

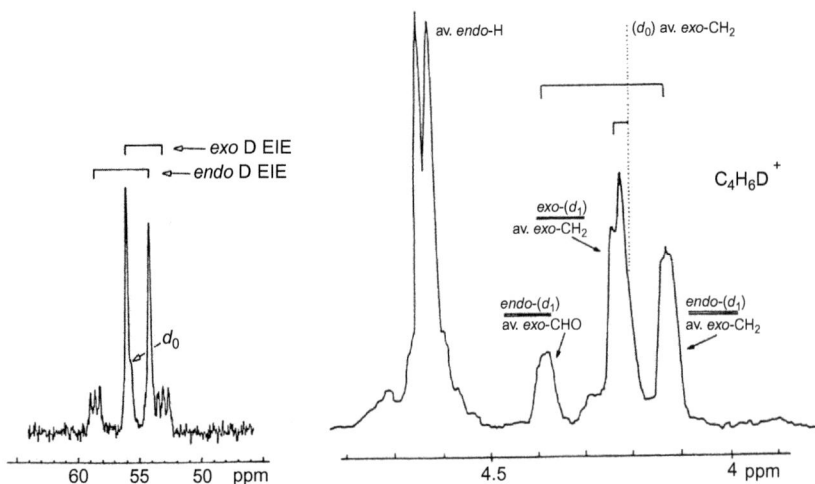

Fig. 7 (A) ^{13}C and (B) ^1H NMR spectra of an *exo*– and *endo*-$(C_4H_6D_1)^+/(C_4H_7)^+$ cation mixture.

averaged methylene groups of $(C_4H_6D)^+$ cations relative to the averaged CH_2 signal in the nondeuterated cation shows that the equilibration process is between two different sites, one singly and one doubly populated.

In the ^1H NMR spectrum (Fig. 7B), the endo-$(C_4H_5D_1)^+$ cation gives rise to a large EIE splitting for the averaged exo-hydrogens. The peak for the exo-hydrogens of the two CH_2-groups with intensity 2 is shielded, and the signal with intensity 1 for the CHD-exo-hydrogen is deshielded twice as much relative to the d_0 cation. The exo-$(C_4H_5D)^+$ cation shows an inverse and much smaller effect. The two remaining exo-hydrogens are somewhat deshielded compared to the d_0 cation. As with the CD_2-labeled cation, the endo-hydrogens of the d_1 cations show no significant isotope effect splittings.

The ^{13}C NMR signal (Fig. 7A) of the averaged methylene groups confirms that the two temperature-dependent EIE splittings are different in sign and magnitude for the sterically distinct exo- and endo-CHD-deuterated $C_4H_6D_1^+$ cations. The chemical shifts of the signals of the deuterated cation are to be corrected for intrinsic deuterium isotope effects on ^{13}C NMR chemical shifts. The endo-d_1 cation then shows for the averaged methylene signals a large isotope splitting with the deuterated methylene carbon triplet signal twice as much deshielded as the two CH_2-groups are shielded compared to the d_0 cation. The exo-d_1 cation shows smaller EIE splittings, the CHD-carbon triplet peak is shielded and the CH_2 signals are somewhat deshielded with respect to the averaged methylene peaks in the d_0 cation.

1.9 Quantum-Chemical Calculations of EIEs in $(C_4H_7)^+$ Cations

The experimental EIE results in deuterated $(C_4H_7)^+$ cations are in full agreement with theoretical calculations. Diagonal force constants for stretches and bending of the methylene C—H bond, reduced isotopic partition functions and equilibrium constants for deuterium isotope effects for the fast methylene rearrangement in bicyclobutonium **9**, and the bisected cyclopropylmethyl cations **7** have been calculated by quantum-chemical calculations.[38] The largest differences are found for the force constants of the C—H bonds at the pentacoordinated carbon Cγ. The exo-Cγ—H bond has a significantly higher force constant than the endo-Cγ—H bond. The exo-C—H bonds at the tetracoordinated carbons $C_\beta,C_{\beta'}$ are less stiff than the exo-Cγ—H bond. The endo-oriented C—H bonds at the $C_\beta,C_{\beta'}$ carbons have higher force constants than the exo-Cγ—H bond. The relative force constants for the C—H bonds in the bicyclobutonium ion are

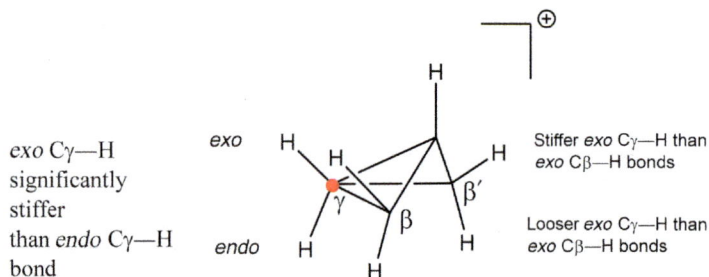

exo Cγ—H significantly stiffer than *endo* Cγ—H bond

exo

Stiffer *exo* Cγ—H than *exo* Cβ—H bonds

Looser *exo* Cγ—H than *exo* Cβ—H bonds

endo

Fig. 8 Relative force constants in cation **9**.[38]

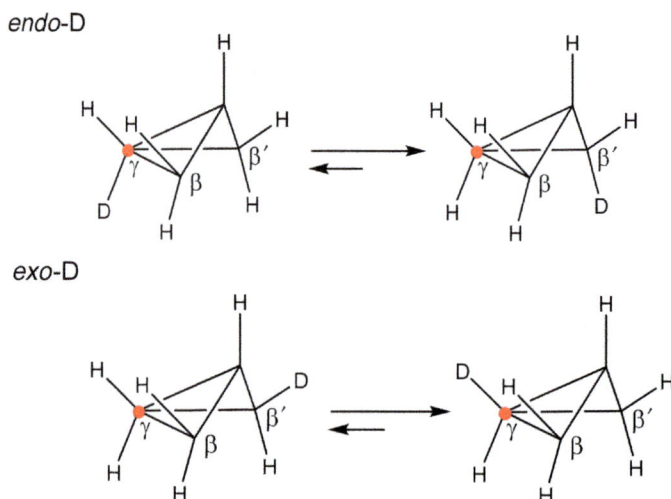

endo-D

exo-D

Fig. 9 Equilibrium perturbation in bicyclobutonium ions **9** by *endo*- and *exo*-deuterium.

depicted in Fig. 8. No such differences are observed for *endo*- and *exo*-deuterated cyclopropylmethyl cations **7**. The calculated direction of the equilibrium perturbation for *endo*- and *exo*-CHD-deuterated bicyclobutonium ions **9** is shown in Fig. 9.

^{1}H and ^{13}C NMR chemical shifts for pristine static structures of the bicyclobutonium ion **9** and the bisected cyclopropylmethyl cation **7** recently have been recalculated with high accuracy by quantum-chemical calculations (CCSD/cc-pVTZ//CCSD(T)/qz2p)[30] and are shown in Figs. 10 and 5, respectively.

1.9.1 The Size and the Direction of the EIEs

The size and the direction of the experimentally observed deuterium EIEs on NMR spectra of deuterated $(C_4H_7)^+$ cations can be explained

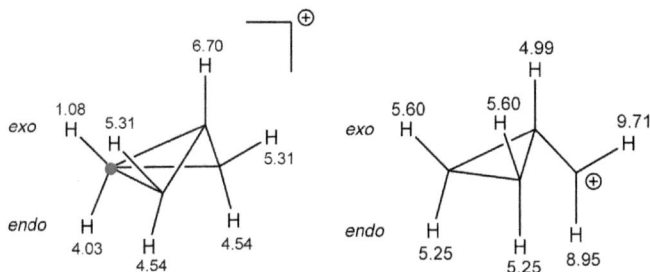

Fig. 10 (CCSD/cc-pVTZ//CCSD(T)/qz2p) calculated ^{1}H NMR chemical shifts of bicyclobutonium ion **9** and cyclopropylmethyl cation **7**.

consistently and unequivocally considering the calculated force constant differences of the C—H bonds of the exchanging sites, and the ab initio calculated chemical shift differences, in accordance with the "first law" of isotope effect chemistry: in an H/D exchange situation between C—H bonding sites with different stiffness, the heavier isotope prefers to make the stiffer bond.

1.9.2 Population of the Equilibrating Sites

The singly populated site of the two-site exchange equilibrium is attributed to the pentacoordinated C_γ carbon with the two hydrogens with very different C—H bond force constants. This gives rise to two isotope effects different in sign and magnitude for the sterically distinct *exo-* and *endo-* CHD-deuterated $(C_4H_6D_1)^+$ cations. The double-populated site is attributed to the two tetracoordinated carbons C_β, $C_{\beta'}$.

1.9.3 The Size of the EIE

The much larger EIE-induced splitting of the *exo*-hydrogen chemical shifts caused by the *endo-d_1* cation compared to the small *exo-d_1* cation-induced deshielding of the two remaining *exo*-hydrogens (Fig. 7B) is due to the calculated large difference in chemical shift, 1.08 vs 5.01 ppm, between the *exo*-C_γ—H and the *exo*-C_β-hydrogens and the relative small chemical shift difference, 4.03 vs 4.54 ppm, between the C_γ- and C_β-bonded *endo*-hydrogens, respectively.

1.9.4 The EIE-Induced Direction of the NMR Chemical Shifts of the Averaged Methylene Signals

In the *endo*-CHD-deuterated $(C_4H_6D)^+$ cation, the equilibrium is perturbed so that deuterium prefers the stiffer *endo*-C_β—H, $C_{\beta'}$—H bonds at the tetracoordinated $C_\beta, C_{\beta'}$ rather than the looser $endo$-C_γ—H bond. The ^{13}C NMR triplet signal of the CHD-methylene carbon in the $endo$-d_1 cation therefore is deshielded compared to the averaged methylene peaks of the d_0 cation shifted toward the direction of the calculated shift of 75.4 ppm for the C_β, $C_{\beta'}$ carbons in a static cation **9**. The two CH_2 groups in the $endo$-d_1 cation are shielded relative to the d_0 cation toward the direction of the shift of the pentacoordinated C_γ carbon ($\delta C_\gamma = -13.7$ ppm calculated for static cation **9**).

In the ^1H NMR spectrum of the $endo$-CHD-deuterated cation, the signal for the single exo-hydrogen at the CHD group is deshielded toward the position of the hydrogens at the tetracoordinated carbons C_β, $C_{\beta'}$ (δ exo-H_β, $H_{\beta'} = 5.31$ ppm calculated for static cation **9**) because the $endo$-deuterium prefers the stiffer bonding position at the C_β, $C_{\beta'}$ carbons. The signal for the two exo-CH_2 hydrogens is deshielded compared to the d_0 cation, which is toward the direction of the shift of the exo-hydrogen at the pentacoordinated C_γ carbon (δ exo-$H\gamma = 1.08$ ppm calculated for static cation **9**), which has the looser exo-C—H bond. The bond to the exo-hydrogen (calc. $\delta^1H = 1.08$ ppm) at the pentacoordinated carbon has a larger force constant than the bond to the exo-hydrogens at the tetracoordinated carbons (calc. $\delta^1H = 5.31$ ppm).

In the exo-CHD-deuterated $(C_4H_6D)^+$ cation, the equilibrium is shifted so that deuterium is preferred at the exo-position on $C\gamma$ which has the stiffest exo-C—H bond. The ^{13}C NMR signal of the deuterated methylene carbon triplet in the exo-CHD-deuterated $C_4H_6D_1{}^+$ cation (Fig. 7A) is shielded relative to the d_0 cation toward the direction of the chemical shift of the pentacoordinated $C\gamma$ carbon (-13.7 ppm calc. for a static bicyclobutonium cation). The ^1H NMR signal of the two remaining exo-hydrogens in exo-deuterated $C_4H_6D_1{}^+$ is somewhat deshielded with respect to the d_0 cation, shifted toward the calculated shift position of the exo-hydrogens at the tetracoordinated methylene carbons (Fig. 7B).

No reasonable structure for $(C_4H_7)^+$ giving EIEs of opposite directions for exo- and $endo$-deuterated isomers was found other than the bicyclobutonium cation **9**.

The interpretation of experimental results on deuterium EIEs on NMR spectra of mono-, di-, and trideuterated cations[36–39] is in accord with the major species having a bridged $(C_4H_7)^+$ bicyclobutonium structure **9** with C_s point group symmetry; there is no evidence for unsymmetrical bicyclobutonium ions such as **14**.

The experimental and computational EIE results clearly prefer the bridged hypercoordinated bicyclobutonium ion structure **9** in the equilibrium of the $(C_4H_7)^+$ cation system and confirm an ingenious forward-looking interpretation made by Olah et al. in 1972 long before ^{13}C FT NMR and ab initio MO methods were available.[7] The stabilization mode of the parent $(C_4H_7)^+$ cation as depicted by structure **12** was correctly anticipated to arise from the bridging interaction of the backside lobe of the $C\gamma$—H_{endo} bond orbital with the formally empty p-orbital at the carbenium carbon at C_α. This early hypothesis is fully supported by the calculations of Saunders and Wolfsberg[38] in 1989, who showed that the *endo*-C—H bond at the pentacoordinated carbon has the lowest bond force constant of all methylene C—H bonds in **9**. Comparable to the origin of the β-deuterium isotope effect observed for β-hyperconjugation in carbocations, a γ-deuterium isotope effect is observed in **9** because electron density is drained from the *endo*-$C\gamma$—H bond toward the formally electron-deficient center at C_α thus leading to a bridging bonding interaction between C_γ and C_α and the reduced bond force constants for the *endo*-C_γ—H bond.

The ^{13}C NMR spectra of d_1- and d_3-deuterated methylene deuterated $(C_4H_7)^+$ cations show a barely temperature-dependent small splitting of ~0.4 ppm for the signal of the methine carbon, which was suggested to indicate an isotope effect upon the nondegenerate equilibrium between the major bicyclobutonium ion **9** and the minor isomeric cyclopropylmethyl cation **7**.[36,37] This idea is also supported by ab initio MO calculations of EIEs for the $(C_4H_7)^+$ cation system.[38]

1.9.5 Quantum-Chemical Calculations for the $(C_4H_7)^+$ Cation System

The flat PES and the small energy differences between different structures of $(C_4H_7)^+$ cations make confident quantum-chemical predictions of the predominant species difficult because the energy differences are within, or close to, the error limit of the calculations.

Numerous calculations have been published over the years with different results.[40] Early quantum-chemical calculations were not decisive. Computational results obtained using DFT methods are in contrast to calculations using wave function methods including a high level of electron correlation, such as MP4 and CCSD methods, and large basis sets.

Different calculated solvent effects for different cyclic isomers of $(C_4H_7)^+$ cations have been reported[15]; however, these effects are rather small.

Recent quantum-chemical calculations including electron correlation for pristine $(C_4H_7)^+$ isomers describe the system more accurately and favor the bicyclobutonium ion by 1–2.4 kcal/mol.[41] CCSD/cc-pVTZ optimized structures and CCSD-f12/aug-ccpVTZ single-point-energies favor the bicyclobutonium ion by $\sim$1 kcal/mol.[30]

The validity and adequacy of quantum-chemical calculations of structure and energy can often be evaluated by comparison of calculated properties such as NMR chemical shifts with experimentally accessible NMR data of static carbocations.[42]

For the parent $(C_4H_7)^+$ cation, precise experimental NMR data for a static structure are not available as the slow exchange limit is not accessible in solution. However, the combination of experimental and computational methods finally helped to solve the conundrum of the $(C_4H_7)^+$ cation system, which is now best described as a triply degenerate set of rapidly interconverting bicyclobutonium ions 9 of C_s point group symmetry with minor contributions from a triply degenerate set of rapidly equilibrating bisected cyclopropylmethyl cations 7 which are only marginally higher in energy.

2. SUBSTITUTED $(C_4H_7)^+$ CARBOCATIONS

The structure and stability of substituted $(C_4H_7)^+$ cations depend on the nature and the position of the substituent.

2.1 Fast-Equilibrating Substituted Bicyclobutonium Ions
2.1.1 Solvolysis Results of Substituted Cyclobutyl Derivatives
Separating transition state effects and product-forming effects is a common problem in the interpretation of substituent effects in solvolysis reactions. An investigation of the acetolysis of 3-substituted cyclobutyl tosylates 15 accompanied by quantum-chemical MO calculations at the MP2/6–31G (d) theoretical level indicated that the rate-determining step involves a transition state which has essentially a bridged cyclobutyl cation structure 16 (bicyclobutonium ion), and that rearrangements to cyclopropylmethyl 17 and homoallyl ions 18 take place only after this step (Fig. 11).[43]

Solvolysis of 3-alkyl- and 3-aryl-substituted cyclobutyl tosylates leads in the rate-determining step to a bridged cyclobutyl cation which then rearranges to the corresponding cyclopropylmethyl cation to give products.[44]

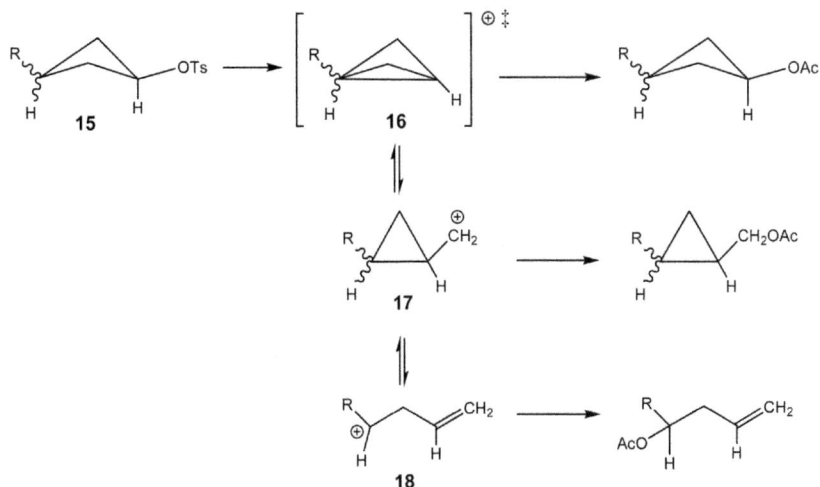

Fig. 11 Acetolysis of 3-substituted cyclobutyl tosylates **15**.

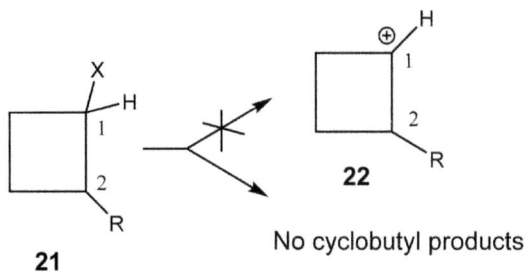

No cyclobutyl products

21

Fig. 12 Solvolysis of 2-substituted cyclobutyl substrates **21**.

1-Substituted cyclobutyl cations $(1-RC_4H_6)^+$ may have the structure of a tricoordinated cyclobutyl cation **19** or a bicyclobutonium ion structure **20** with a pentacoordinated C_γ carbon.

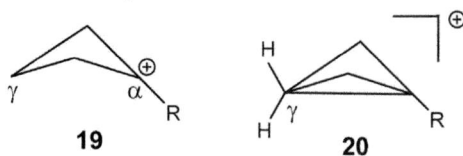

2-Substituted cyclobutyl substrates **21** generally give only cyclopropylmethyl derivatives in solvolysis and deamination reactions (Fig. 12).[45] Thus 2-substituted cyclobutyl cations **22** have not been considered as short-lived intermediates and also could not be generated in superacid media.[46]

2.2 NMR Spectroscopic and Computational Investigations of Substituted $(C_4H_7)^+$ Cations

2.2.1 1-Methyl-Bicyclobutonium Cation

Similar to the parent $(C_4H_7)^+$ cation, the dominant structure in the dynamic equilibrium of the 1-methylcyclobutyl/1-methylcyclopropylmethyl cation $(C_4H_6CH_3)^+$ has been in dispute for a long time.[7,47–50]

1-Methylcyclobutyl- (**23**), 1′-methylcyclopropylmethyl- (**24**), and 1-methylbicyclobutonium cation (**25**) structures have been suggested.

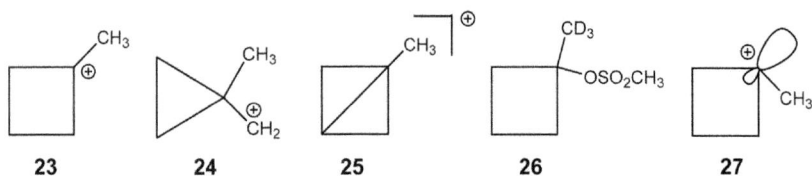

23   **24**   **25**   **26**   **27**

The reduced magnitude of the secondary deuterium isotope effects in the solvolysis of 1-CD_3-cyclobutyl methanesulfonate **26** was suggested to originate from a transition state structurally closely related to an intermediate 1-methylbicyclobutonium ion (**25**-CD_3).[43]

On the basis of ^{13}C NMR spectroscopic studies, two different structures have been suggested. One is an sp^3-hybridized cyclobutyl cation **27** and the other is a bicyclobutonium **25**, which is considered to be either a single minimum **28** or a set of fast-equilibrating less symmetric cations **29a** and **29b** claimed to be indistinguishable from a symmetric one.[49,50] A nondegenerate equilibrium of bicyclobutonium ions and 1′-methylcyclopropylmethyl-cations as a minor isomer has also been discussed.[47,51]

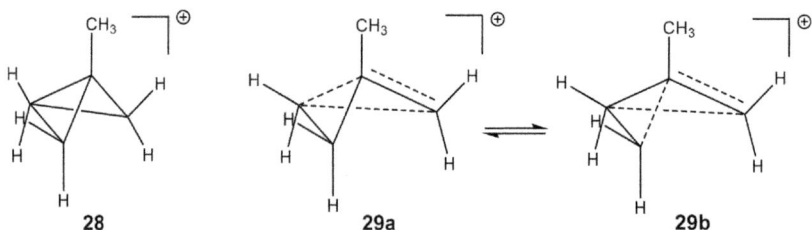

28   **29a**   **29b**

A comparison of the deuterium EIEs in the NMR spectra of deuterated $(C_4H_6CH_3)^+$ cations with the EIE in the deuterated parent $(C_4H_7)^+$ cation and in related cyclopropylmethyl- and cyclobutyl-type model cations could clarify the controversy concerning the structure of $(C_4H_6CH_3)^+$ cation.[39,52,53]

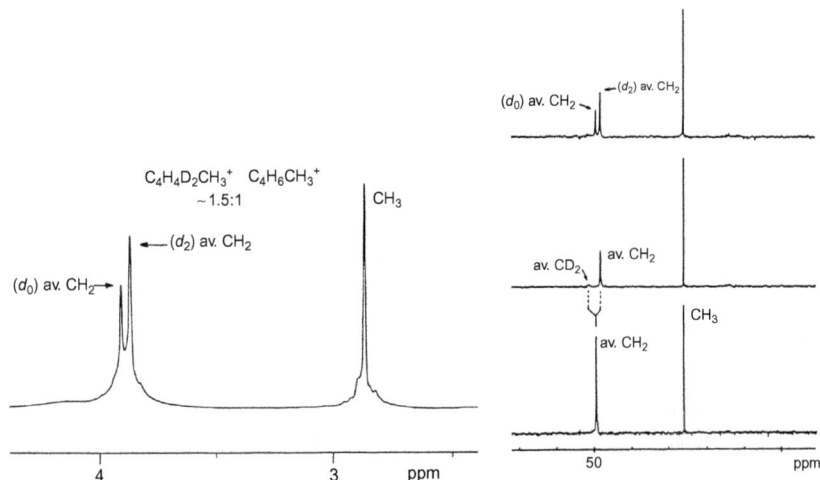

Fig. 13 (A) ^{1}H NMR of $(C_4H_4D_2CH_3)^+/(C_4H_5CH_3)^+$ mixture. (B) Comparison of ^{13}C NMR spectra *bottom*: $(C_4H_5CH_3)^+$, *middle*: $(C_4H_4D_2CH_3)^+$, and *top*: mixture of $(C_4H_4D_2CH_3)^+$ and $(C_4H_5CH_3)^+$.

The ^{1}H and ^{13}C NMR spectra of a mixture of unlabeled and CD_2-labeled $(C_4H_4D_2CH_3)^+$ show isotopic perturbation for the averaged signal of the three methylene groups, whereas the other peaks are unaffected (Fig. 13).

The peak for the nondeuterated methylene groups in $C_4H_4D_2CH_3{}^+$ show temperature-dependent shielding in the ^{1}H and ^{13}C NMR spectra compared to the unlabeled ion (Fig. 13). The CD_2-carbon signal is deshielded (Fig. 13B, middle trace). These spectra indicate a definite EIE on the fast methylene rearrangement.

In the CHD-monolabeled $(C_4H_5DCH_3)^+$ cation, the signal for the methylene proton geminal to deuterium is deshielded (Fig. 14A) whereas the remaining methylene protons are shielded relative to the peak of the d_0 cation.

The ^{13}C NMR spectrum of a mixture of $(C_4H_5DCH_3)^+$ and $(C_4H_6CH_3)^+$ (Fig. 14B, middle trace) shows the triplet of the d_1-deuterated methylene carbon deshielded and the signal for the two CH_2 carbons in $(C_4H_6DCH_3)^+$ shielded compared to the shift of averaged CH_2 carbons in the unlabeled cation. The deshielding of the d_1-deuterated methylene carbon relative to the d_0 cation, corrected for an intrinsic shift, has twice the size of the shielding of the CH_2-signal, indicating a two-site fast exchange taking place between a double-populated deshielded and a single-populated shielded site.

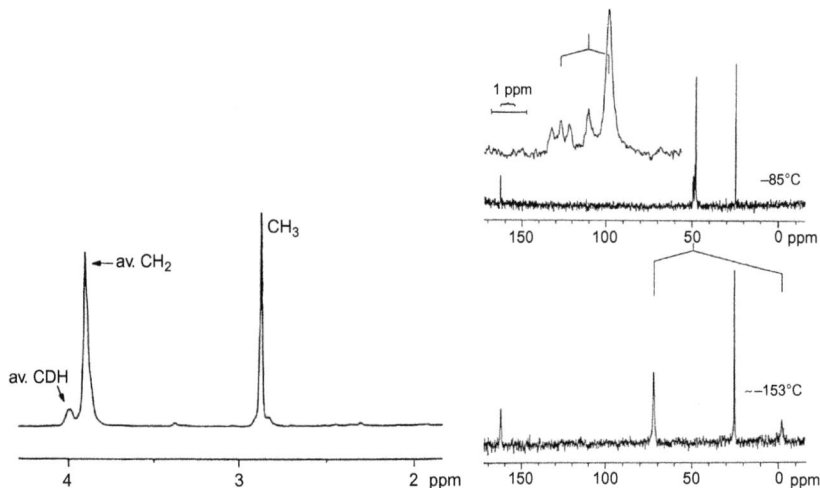

Fig. 14 (A) ^{1}H NMR of $(C_4H_5DCH_3)^+$ and (B) *top*: ^{13}C NMR spectra of $(C_4H_5DCH_3)^+$ admixed with some $(C_4H_6CH_3)^+$, *bottom*: slow exchange ^{13}C NMR spectra of $(C_4H_6CH_3)^+$ at -153°C.

At about -153°C the equilibrium is at or close to the slow exchange limit (Fig. 14B, bottom trace). The methylene carbons are decoalesced and separated into two broad peaks at 71.3 and -2.8 ppm (ratio 2:1). The equilibrium averaging the methylene groups is frozen out and consequently the isotopic perturbation of this process is no longer observable.

Contrary to the parent $(C_4H_7)^+$ cation, the geminal hydrogens at the averaged methylene groups of $(C_4H_6CH_3)^+$ are not distinct and give only one signal because of conformational averaging. The 1-methylbicyclobutonium cation **28** undergoes a ring inversion process via a planar cyclobutyl cation **32** (R$=$CH$_3$) transition state averaging the geminal-related *exo-* and *endo*-hydrogens at the methylene groups (Fig. 15). In the 1–CH$_3$-substituted bicyclobutonium ion **28**, the nonbridged puckered and planar cyclobutyl cation structures **31** and **32** (R$=$CH$_3$) are lower in energy than for the parent bicyclobutonium ion **9** (**31**, **32**, R$=$H) due to better stabilization of the positive charge by an α-CH$_3$ group as compared to an α-H.

In the course of this conformational averaging, bridged 1-methylbicyclobutonium ion structures **30a**, **30b**, and **30c** (R$=$CH$_3$) change to puckered 1-methylcyclobutyl cations such as **31a** (R$=$CH$_3$), which undergo ring inversion to **31b** via a planar 1-methylcyclobutyl structure **32** (R$=$CH$_3$) to form a set of mirror images **58d**, **58e**, and **58f** (R$=$CH$_3$).

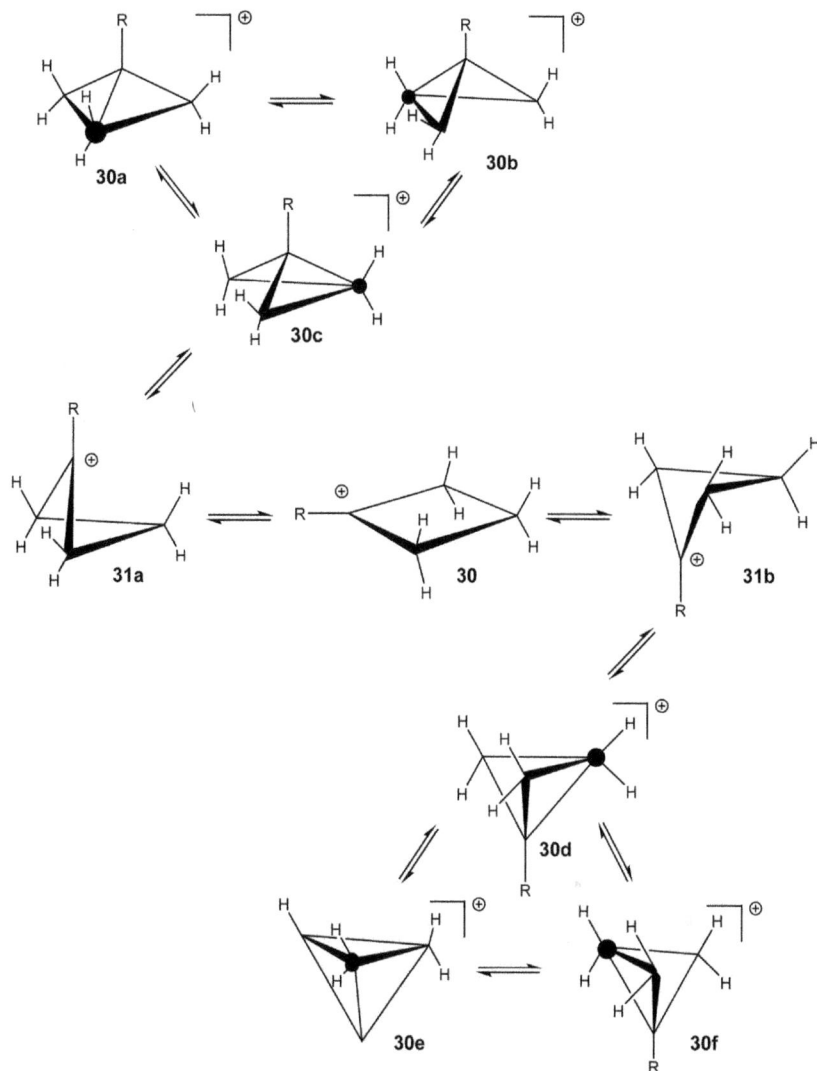

Fig. 15 Methylene rearrangement and conformational averaging of $(C_4H_6R)^+$ cations.

The direction of the EIE in the ^{13}C NMR spectra of **28** indicates that the mean of the C—H bond force constant at the single methylene group, which gives the signal at −2.8 ppm in the static cation at −153°C, must be lower than the C—H bond force constants at the two methylene carbons which give rise to the signal at 71.3 ppm at −153°C. Contrary to the parent $(C_4H_7)^+$ cation, the signals of the $(C_4H_6CH_3)^+$ cation are not significantly temperature

c. 24
29.24
CH_3 ⊕

c. 163
154.5

H
H
H

−2.8
−2.9

71.3

71.3   76.3
76.3

H

H

H

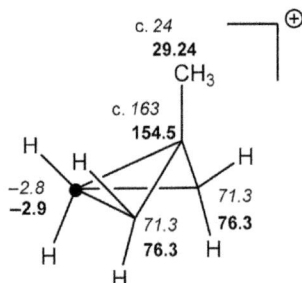

Fig. 16 *Italic*: experimental ^{13}C NMR chemical shifts of **28** at c. −153°C; *bold*: mp2/cc-pVTZ//mp2/cc-pVTZ calculated ^{13}C NMR chemical shifts of **28**.

dependent. A minor isomer such as the isomeric 1′-methylcyclopropylmethyl cation structure **24** does not contribute to the averaged chemical shifts. Structure **24** is calculated (MP2//cc-pvtz) to be a TS.[54] The calculated ^{13}NMR chemical shifts for a pristine static 1-methylbicyclobutonium ion are in reasonable agreement with the experimentally observed chemical shifts at −153°C (Fig. 16).

2.2.2 The 1-(Trimethylsilyl)Bicyclobutonium Ion

An NMR spectroscopic and computational study of the α-silyl effect in benzyl-type cations has shown that a α-trimethylsilyl group stabilizes a positive charge less than an α-methyl group but better than an α-hydrogen (Fig. 17).[55]

The 1-(trimethylsilyl)bicyclobutonium ion **33** is formed exclusively when (1-(trimethylsilyl)cyclopropyl)methanol **34** is reacted with SbF_5. The isomeric 1-trimethylsilylcyclopropylmethyl cation **35** is not formed (Fig. 18).[56]

The ^{13}C and ^1H NMR spectroscopic data show that **33** has a bridged puckered bicyclobutonium structure and undergo a rapid threefold-degenerate rearrangement as shown for **30a**, **30b**, and **30c** [R = Si(CH_3)_3] that renders the two β- and one γ-methylene groups equivalent, leading to one averaged ^{13}C NMR signal for the CH_2-groups at 48.9 ppm (Fig. 19).

Kinetic line broadening is not observed at temperatures as low as −130°C.

The ^{13}C chemical shifts of cation **33** show only a negligible temperature dependence in contrast to the ^{13}C spectra of the parent $(C_4H_7)^+$ cations. Thus a contribution of a minor isomer in the dynamic equilibrium of **33** is not very likely. The conceivable isomeric 1′-trimethylsilylcyclopropylmethyl cation **35** is calculated to be a TS (MP2/cc-p-VTZ).

Like in the parent bicyclobutonium ion **9** [**30**, R = H] conformational ring inversion for **33** [**30**, R = Si(CH_3)_3] is slow, so that two separate signals

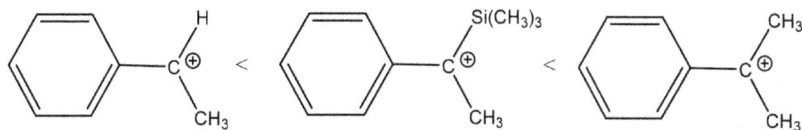

Fig. 17 Stabilization of positive charge in substituted benzyl-type cations.

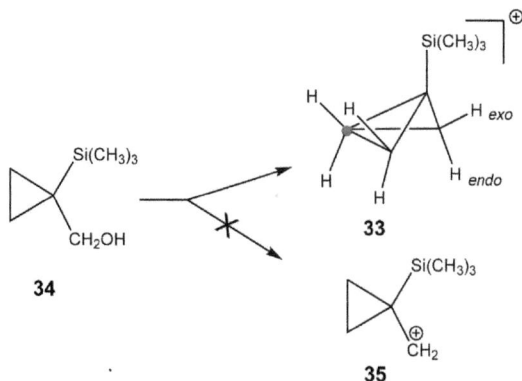

Fig. 18 Formation of 1-(trimethylsilyl)bicyclobutonium ion **33**.

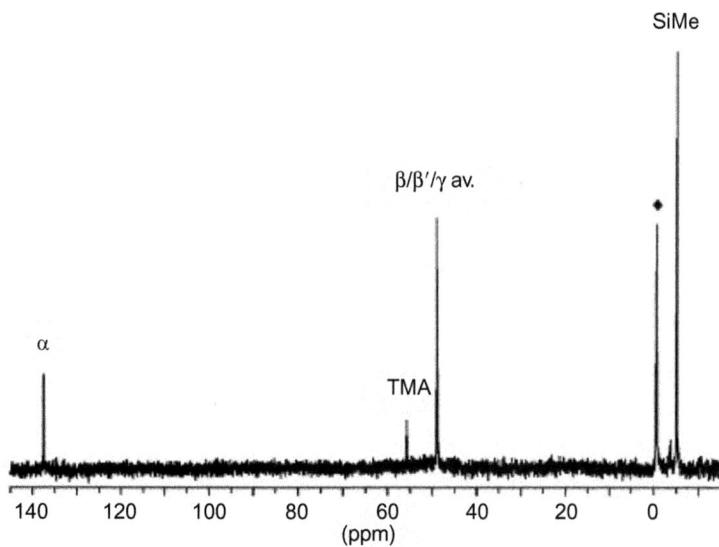

Fig. 19 100 MHz ^{13}C NMR spectrum of the 1-(trimethylsilyl)bicyclobutonium ion **33** (◆: FSiMe$_3$) at $-128°C$ (internal standard TMA $\delta(NMe_4{}^+) = 55.65$ ppm).

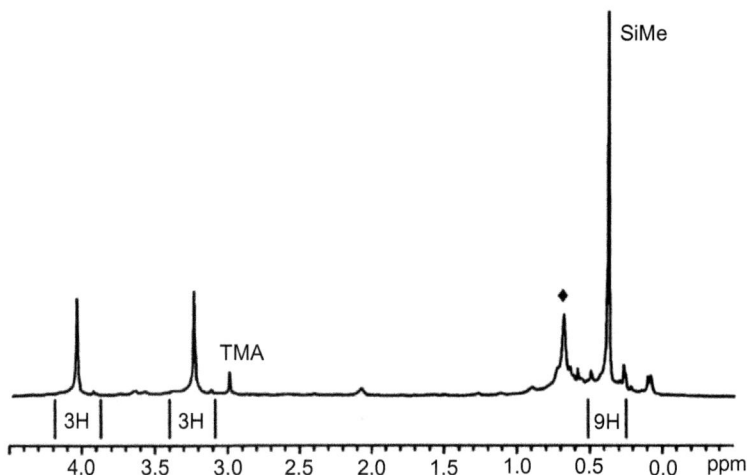

Fig. 20 ^{1}H NMR spectrum of 1-(trimethylsilyl)bicyclobutonium ion **33**.

for the three averaged endo-CH_2 (4.04 ppm) and three averaged exo-CH_2 hydrogens (3.24 ppm) are observed (Fig. 20).

The exo/endo equilibration via a planar TS for ring inversion [**32**, R=H or Si$(CH_3)_3$] (Fig. 15) is energetically not accessible at temperatures where these cations are stable. In the case of the 1-CH_3-substituted bicyclo-butonium ion **28** (**30**, R=CH_3), a nonbridged puckered cyclobutyl and a planar cyclobutyl cation structures **31** and **32** (R=CH_3) are lower in energy due to better stabilization of the positive charge by an α-CH_3 group as compared to an α-H or α-Si$(CH_3)_3$-group.

The ^{1}H NMR spectrum of a mixture exo- and endo-CHD-monolabeled **33** and **33-d_0** (Fig. 21) shows deuterium EIEs which are different in sign and magnitude and are rationalized by different endo- and exo-C—H bond force constants at the pentacoordinated carbon, similar to the parent bicyclo-butonium cation **9** (compare Figs. 21 and 7B). Fig. 22 shows the different EIEs of exo- and endo-CHD-monolabeled **33** on the averaged methylene peak in the ^{13}C NMR spectrum which are analogous the those in the CHD-monolabeled parent cation (compare Figs. 22 and 7A).

^{13}C NMR chemical shift calculations for the 1-trimethylsilylbicy-clobutonium ion **33**, with averaged chemical shift for the three CH_2 groups, satisfactorily reproduce the experimentally observed chemical shifts (Fig. 23).

The geometric and electronic properties of the 1-(trimethylsilyl)-bicyclobutonium cation **33** are intermediate between those of the parent

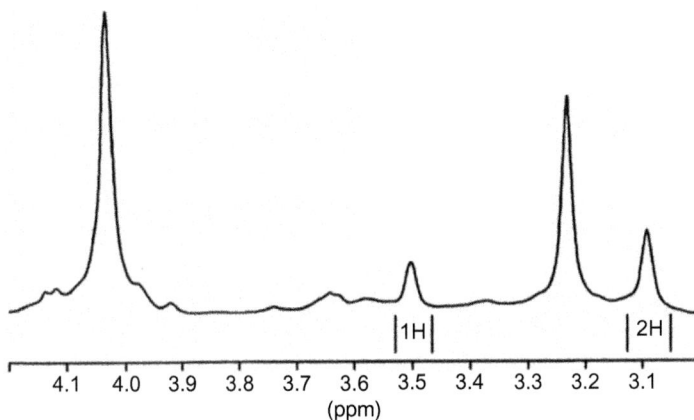

Fig. 21 ^{1}H NMR spectrum of a mixture *exo*- and *endo*-CHD-monolabeled monolabeled **33** and **33-d_0**.

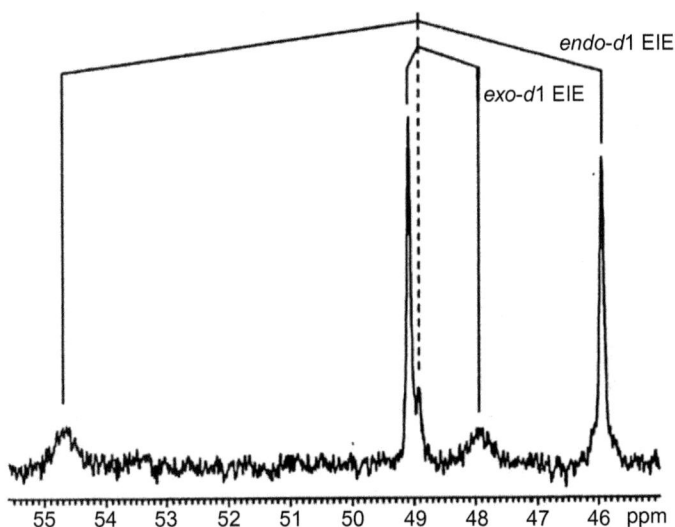

Fig. 22 Methylene region of ^{13}C NMR spectrum of a mixture *exo*- and *endo*-CHD-monolabeled **33** admixed with some **33-d_0**.

bicyclobutonium ion **9** and the methyl-substituted bicyclobutonium ion **28**. Similar to **28**, no other isomer contributes to the observed NMR spectra of **33**. In contrast to the methyl-substituted cation **28**, but analogous to the parent bicyclobutoniumion **9**, cation **33** undergoes no conformational ring inversion.

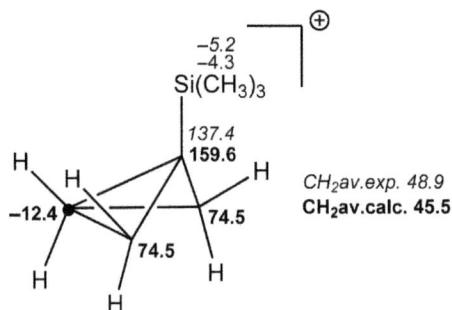

Fig. 23 *Italic*: experimental ^{13}C NMR chemical shifts of **33**; *bold*: mp2/cc-pVTZ//mp2/ cc-pVTZ calculated ^{13}C NMR chemical shifts of **33**.

2.3 EIEs in Substituted Cyclopropylmethyl and Substituted Cyclobutyl Cations

Observed deuterium EIEs have been interpreted as arising either predominantly (for the parent $C_4H_7^+$ cation) or exclusively, for the 1-substituted $C_4H_6R^+$ cations **28** (R=CH$_3$) and **33** (R=Si(CH$_3$)$_3$), from a hypercoordinated bicyclobutonium structure. This is supported by deuterium EIEs of suitable model cations for the suggested alternative cyclopropylmethyl- and cyclobutyl-type structures **7, 8, 23, 24**, and **27**. For the cyclopropyl-type $(C_4H_6R)^+$ cation structures considered (**7, 24**, and **35**), suitable model structures are methyl-substituted cyclopropylmethyl cations, whose structure and dynamics are not in dispute, such as **36, 37**, and **38** which, respectively, undergo twofold- (**36, 38**) and threefold- (**37**) degenerate cyclopropyl–cyclopropyl cation rearrangements leading to averaged signals of the exchanging sites in the ^{1}H and ^{13}C NMR spectra under conditions of fast exchange.

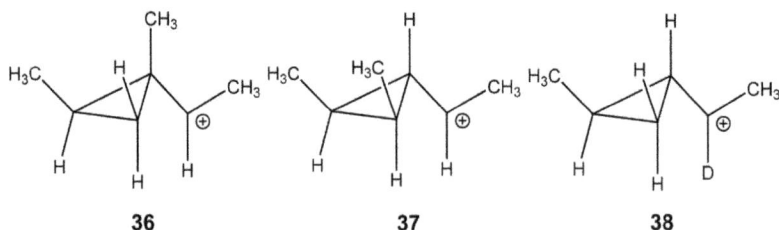

36   **37**   **38**

Fig. 24 shows the temperature-dependent EIE splittings observed for the ^{13}C NMR signal of the averaged methine carbons in the methine monodeuteriated 1-(1,2-dimethylcyclopropyl)ethyl cation d_1-**36**.[53,57]

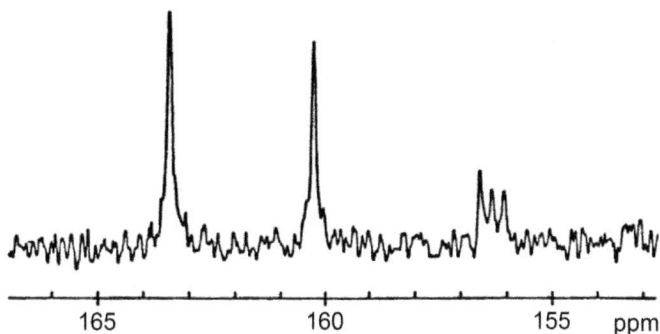

Fig. 24 Methine region of ^{13}C NMR spectrum of d_1-**36** and d_0-**36**.

The splitting of the signal for the averaged methine carbons of d_1-**36**, corrected for an intrinsic isotope shift for the triplet of the deuterated methine carbon, is symmetric with respect to the shift of the carbon of the averaged methine signal in the unlabeled cation **36**, as expected for a two-site exchange between two singly populated sites. The triplet signal of the deuterated methine carbon of d_1-**36** is shielded compared to the signal of the d_0 cation. This is the inverse direction as compared to d_1-methine labeled bicyclobutonium ion **28** (compare Figs. 24 and 14B, top trace), thus excluding a cyclopropylmethyl-type structure **24** for the $(C_4H_6CH_3)^+$ cation. Consistently for the parent $(C_4H_7)^+$ cation, a cyclopropylmethyl-type structure **7** can be excluded as being the major isomer in favor of the bicyclobutonium ion structure **9**. Analogously for the trimethylsilyl-substituted cation, the hypercoordinated structure **33** is preferred, and the cyclopropylmethyl-type structure **35** is ruled out. The deuterated carbon signal in d_1-**36** is moved toward the chemical shift of the cyclopropyl methine carbon in the slow exchange spectra. C–D vibrations are more confined at the cyclopropyl ring than at the C^+-carbon position.[58] The equilibrium d_1-**36a**/d_1-**36b** is shifted toward d_1-**36b**.

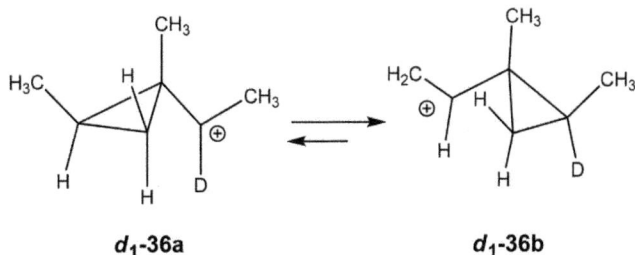

d_1-**36a** d_1-**36b**

^{1}H and ^{13}C NMR spectroscopic investigation of a similar differential α- vs γ-isotope effect for the equilibrium perturbation of the d_1-deuterated 1-(2-methylcyclopropyl)ethyl cation d_1-38a/d_1-38b has also been reported.[59]

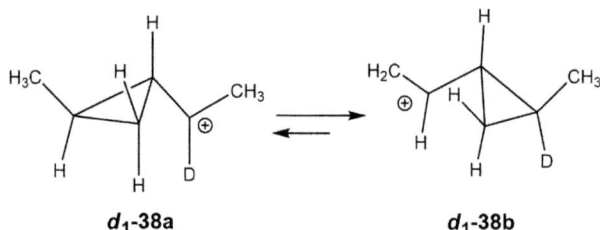

d_1-38a d_1-38b

The 1-(2,3-dimethylcyclopropyl)ethyl cation **37** undergoes at a triply degenerate cyclopropylmethyl/cyclopropylmethyl cation rearrangement which, under conditions of fast exchange, leads to an averaged ^{13}C NMR signal for the three methine carbons and an averaged signal for the three CH–CH$_3$ methyl groups.[60]

At −154°C the fast rearrangement process is slow. In the 100 MHz ^{13}C NMR spectrum, the averaged lines are decoalesced (Fig. 25).[58]

The chemical shifts for the signals under conditions of slow exchange are in accord with a 1-(2,3-dimethylcyclopropyl)ethyl cation with a tri-coordinated cationic carbon center, as assumed from earlier fast exchange

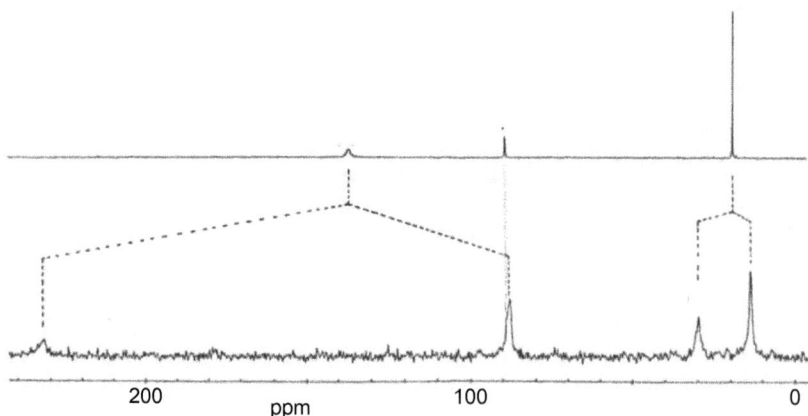

Fig. 25 ^{13}C NMR spectrum of 1-(2,3-dimethylcyclopropyl)ethyl cation **37** at −55°C (*top*) and at −154°C (*bottom*).

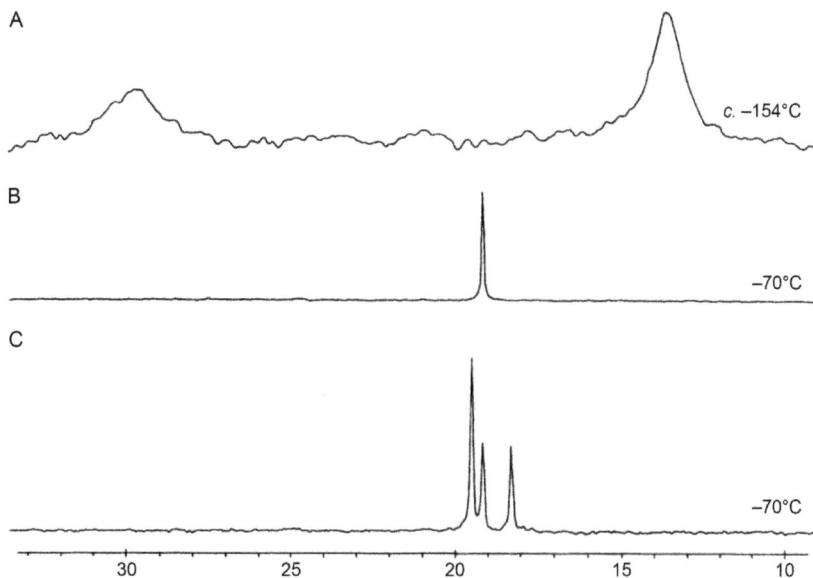

Fig. 26 Methyl region of ^{13}C NMR spectra (A) **37**, 154°C, (B) **37**, −70°C, and (C) 3:1 mixture of **37-d_1** and **37-d_0**, −70°C.

^{13}C NMR spectra[60] and as deduced from applying the additivity of chemical shifts analysis to cation **37**.[61] The ^{13}C NMR spectrum of a mixture of **37** and **d_1-37** under conditions of fast exchange shows characteristic temperature-dependent isotope splittings of the averaged signals for **d_1-37**.[58] Fig. 26 (trace c) shows the EIE splittings of the methyl signals from a 3:1 mixture of **d_1-37** and **37**.

The two outer lines are from the deuterated ion **d_1-37**. Intensity ratio and relative size of the isotope splittings with respect to the middle peak of the d_0 cation **37** indicate that a two-site fast exchange taking place between the singly populated deshielded site (the C^+-CH_3 position) and the doubly populated shielded site (the two equivalent cyclopropyl CH_3 positions) is perturbed by deuterium. The direction of the isotope effect with a peak of relative intensity 1, which shows extra broadening due to unresolved $^2J_{CD}$ spin–spin coupling, is shielded relative to the signal of unlabeled cation, indicating that the deuterated methine group is preferentially on the cyclopropyl ring. The threefold-degenerate equilibrium in **37** in the deuterated cation **d_1-37** is shifted toward **d_1-37b** and **d_1-37c**.

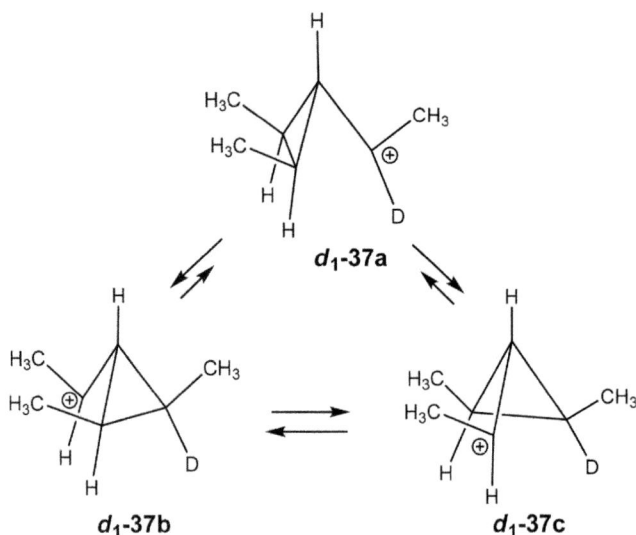

d_1-37a

d_1-37b d_1-37c

The direction of the isotope effect in d_1-37 and the intensity/shift ratio caused by the isotopic perturbation of the triply degenerate rearrangement 1-(2,3-dimethylcyclopropyl)ethyl cation 37, likewise those in d_1-36 and d_1-38, are inverse to those observed for the isotopic perturbation of the triply degenerate rearrangements in the hypercoordinated bicyclobutonium ions $(C_4H_6R)^+$, 9 (R=H), 28 (R=CH$_3$), and 33 (R=Si(CH$_3$)$_3$, thus unequivocally excluding a cyclopropyl-type structure for these cations.

Nonbridged cyclobutyl-type structures 19, such as 8, 23, and 27 for the cations $(C_4H_6R)^+$ (R=H, CH$_3$, and Si(CH$_3$)$_3$ were excluded by comparison of the EIE on NMR spectra of suitable substituted cyclobutyl cations.

1-Aryl-substituted-1-cyclobutyl cations have been characterized as trivalent tertiary cations. Depending on the electron-donating capability of the 1-aryl substituent, they may undergo a threefold-degenerate cyclobutyl/cyclobutyl rearrangement.[62] The 1-[3′,5′-bis(trifluoromethyl)phenyl]cyclobutyl-1-cation 39 is static at −125°C but undergoes a fast threefold-degenerate cyclobutyl/cyclobutyl cation rearrangement at higher temperatures.

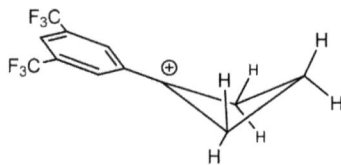

39

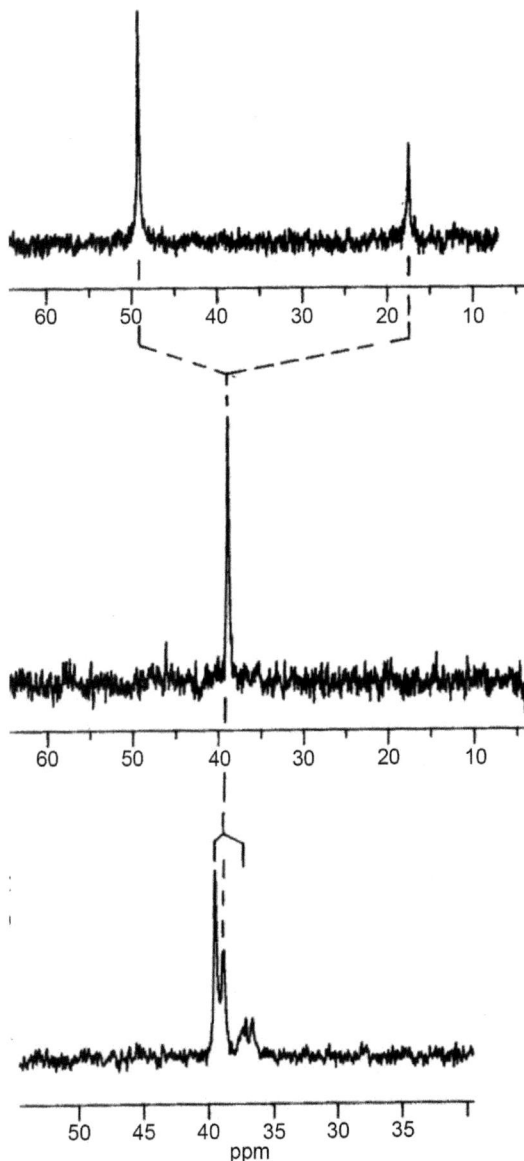

Fig. 27 Methylene region of ^{13}C NMR spectra *top*: **39**–130°C, *middle*: **39**, −37°C, *bottom*: 1:3 mixture of **39**, and d_1-**39**, −37°C.

Under conditions of fast exchange above coalescence, one averaged ^{13}C NMR signal for two β- and one γ-methylene groups is observed (Fig. 27, middle trace). The threefold degeneracy of the equilibrium is perturbed if the symmetry is broken by one or two deuterium in one methylene group.[53,63] Fig. 27 (bottom trace) shows the ^{13}C NMR spectrum of the signals of the averaged methylene groups of a mixture of **39** and d_1**-39**.

The triplet of the CHD group of d_1**-39** is deshielded relative to the central peak of the methylene groups of **39** toward the direction of the γ-methylene signal, whereas the signals for the two CH$_2$-groups in d_1**-39** are deshielded toward the signal for the β-methylene groups in the slow exchange spectrum. Hyperconjugational donation of electron density from the β-C—H bonds to the formal positively charged α-C$^+$-carbon results in lower force constants of the β-bonds compared to the C—H bonds at the γ-position. Deuterium thus is preferred at the γ-carbon and the equilibrium d_1**-39a**/d_1**-39b**/d_1**-39c** is shifted toward d_1**-39a** (Fig. 28).

The ^{1}H NMR spectra of d_1-methylene labeled cation d_1**-39** give proof for the puckered cyclobutyl ring undergoing fast ring inversion. The ^{1}H NMR signal for the deshielded peak of the CH$_2$-protons in d_1**-39** is split into two lines of equal intensity by an additional temperature-dependent isotope effect on top of the EIE on the methylene rearrangement process. This isotope effect perturbs the conformational ring inversion equilibrium. The axial and equatorial β-C—H bonds have different hyperconjugative

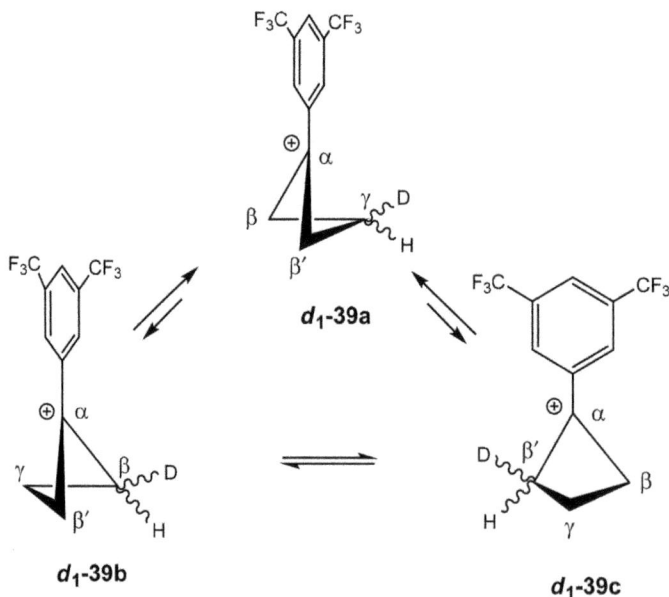

Fig. 28 Equilibrium of mono-CHD deuterated 1-[3′,5′-bis(trifluoromethyl)phenyl]cyclobutyl-1-cation d_1**-39**.

interactions with the formally vacant p-orbital at the α-C^+-carbon and thus different bond force constants. The cyclobutyl ring inversion equilibrium interchanges β-deuterium between axial and equatorial positions, and thus an isotope effect on the ring inversion equilibrium is observed.

A comparison of the EIEs in the ^{13}C NMR spectra of the aryl-substituted cyclobutyl cation $\mathbf{d_1}$-$\mathbf{39}$ (Fig. 27) and the ^{13}C spectra of the methyl-substituted hypercoordinated bridged bicyclobutonium ion $\mathbf{d_1}$-$\mathbf{28}$ (Fig. 14B, top trace) clearly shows that nonbridged cyclobutyl cation structures such as $\mathbf{23}$ and $\mathbf{27}$ can be excluded for the $(C_4H_6CH_3)^+$ cation. This applies analogously for the 1-H- and 1-Si$(CH_3)_3$ substituted bicyclobutonium ions $\mathbf{9}$ and $\mathbf{33}$.

The evidence from experimental isotope effect studies of suitable model cations, without recourse to calculated small energy differences which are almost within uncertainties of the computational methods, is sufficient to characterize the structure and dynamics of the parent $(C_4H_7)^+$ cation—once called a molecular will-o'-the-wisp[29]—and its methyl-substituted analog $(C_4H_6CH_3)^+$, and clarifies the controversy of the minimum-energy structure and dynamics in these cation systems.

2.4 Static Substituted Bicyclobutonium Ions

The stabilization mode of the parent bicyclobutonium ion $\mathbf{9}$ was envisaged correctly by George Olah in 1972, before appropriate methods for MO calculations and FT-^{13}C NMR spectroscopy were established.[7] With an ingenious foresight, he postulated the stabilization of the bicyclobutonium ion to arise from C_α–C_γ bridging by interaction of the backside lobe of the C_γ-H_{endo} orbital with the formally empty carbenium carbon p-orbital at C_α, as depicted in structure $\mathbf{12}$.

If the hydrogen at the $C\gamma$—H_{endo} position is substituted by a better electron–donor group such as a silyl substituent, as shown in structure $\mathbf{40}$, it is to be anticipated that the electron density of the bridging bond is increased, thereby increasing the stability of the bridged hypercoordinated structure and conceivably leading to a static bicyclobutonium ion.

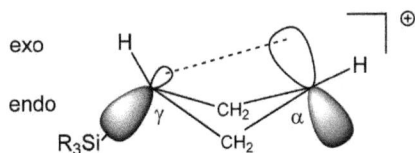

40

The static *endo*-3-trimethlylsilyl-substituted bicyclobutonium ions **41** are directly accessible by ionization of 3-(trimethylsilyl)cyclobutyl chloride **42** with SbF_5 (Fig. 29).[46,64]

The *endo*-3-trimethylsilyl-substituted species **41** is a static bicyclo-butonium ion. The ^{13}C NMR spectrum (Fig. 30) and the H,C–COSY 2D NMR spectra of the corresponding *tert*-butyldimethylsilyl substituted cation **43** (Fig. 31) are a direct proof for the hypercoordinated bicyclobutonium structure of these carbocations.

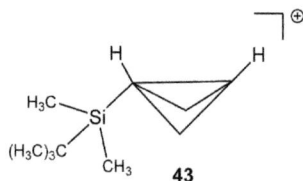

The COSY-45 2D NMR spectrum of **43** shows the coupling correlation of the C^1—H and the strongly shielded C^4—H proton (Fig. 32).

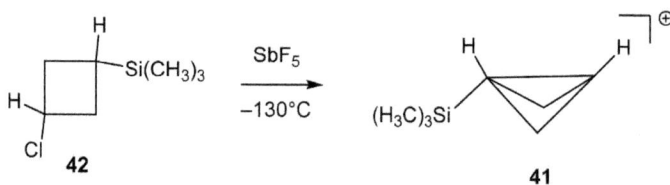

Fig. 29 Generation of *endo*-3-trimethylsilyl-substituted bicyclobutonium ions **41**.

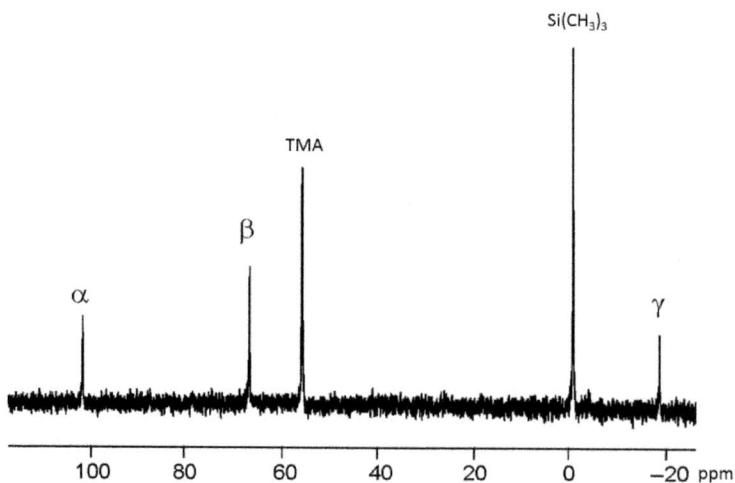

Fig. 30 ^{13}C NMR spectra of *endo*-3-trimethylsilyl-substituted bicyclobutonium ion **41**.

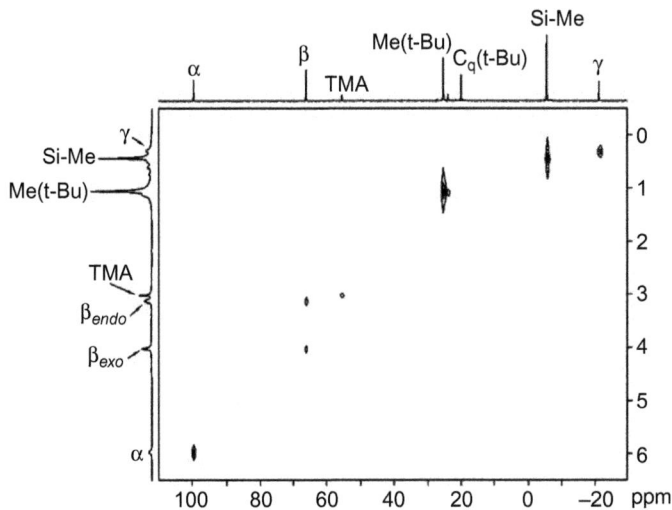

Fig. 31 H,C-COSY 2D NMR spectra of *tert*-butyldimethylsilyl substituted cation **43**.

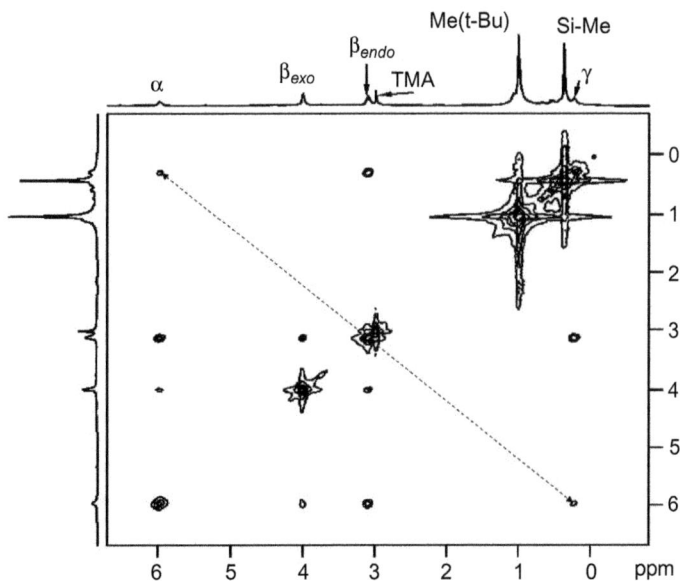

Fig. 32 COSY45 NMR spectra of **43** at −120°C.

The *endo*-orientation of the silyl substituent was confirmed by comparison of the experimentally measured transannular $^3J_{(H,H)}$ spin–spin coupling constant of 5.5 Hz for **43**, which is calculated to be 5.9 Hz for the *endo*-silyl isomer **44** but only 1.2 Hz for the *exo*-silyl isomer **45**.

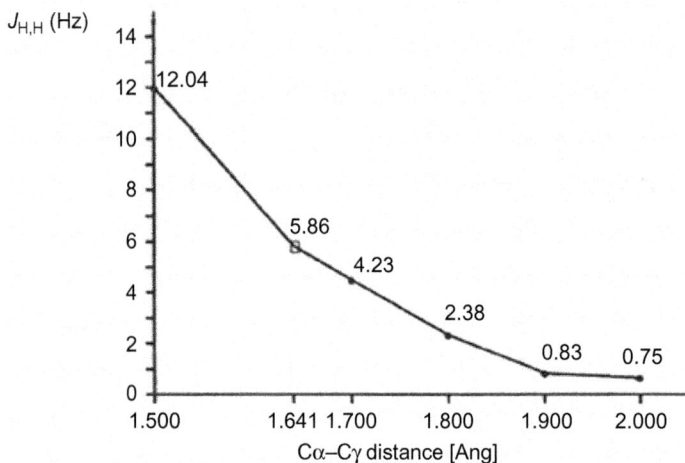

Fig. 33 Bridging bond length dependence of J_{HH} spin–spin coupling between C^1—H and C^4—H in model cation **44**.

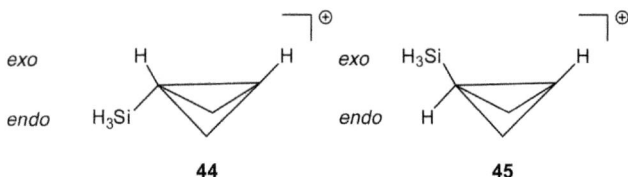

44   **45**

Proof for the vicinal $^3J_{HH}$ coupling pathway between H_α and H_γ across the bridging bond was obtained by calculation of the J_{HH} spin–spin coupling constant as a function of the length of the bridging C_α—C_γ bond. Fig. 33 shows the rapid decline of J_{HH} with increasing C_α–C_γ distance, as expected for a vicinal coupling across the bridging bond, and thus excluding a long-range $^4J_{HH}$ coupling via the C_β carbons.

The structure of the static *endo*-3-trimethylsilylbicyclobutonium ion **41** was finally confirmed by the excellent agreement of the experimental ^{13}C NMR chemical shifts with CCSD/cc-pVTZ//CCSD(T)/qz2p calculated chemical shifts for structure **41** (Fig. 34).

The measured ^{29}Si NMR chemical shift of 43.1 ppm for the *tert*-butyldimethylsilyl substituted bicyclobutonium ion cation **43** is in good agreement with the CCSD/cc-pVTZ//CCSD(T)/qz2p calculated ^{29}Si chemical shift of 49.9 ppm for the trimethylsilyl-substituted bicyclo-butonium ion **41**. The stabilization of the positive charge by the *endo* γ-silyl group in bicyclobutonium ions has a comparable influence on the ^{29}Si chemical shift as observed in the carbocations **46** (56.9 ppm) and **47** (40.7 ppm) which are stabilized by a β-silyl effect.

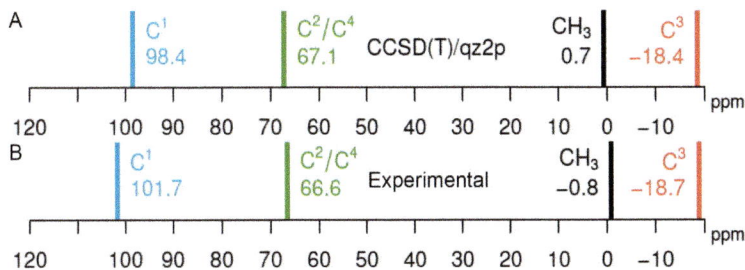

Fig. 34 (A) CCSD/cc-pVTZ//CCSD(T)/qz2p calculated and (B) experimental ^{13}C NMR chemical shifts of *endo*-3-trimethylsilylbicyclobutonium-ion **41**.

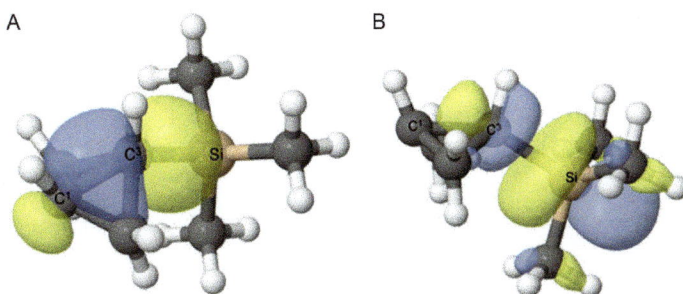

Fig. 35 NLMO calculation of (A) the C1–C3 bridging bond and (B) the lone pair at silicon of the bicyclobutonium ion **41**.

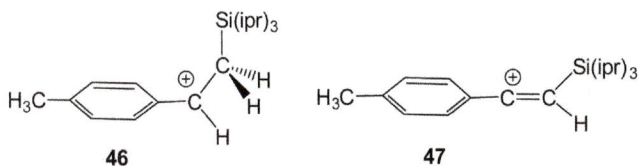

The bridging bond between C_α and C_γ in the 3-*endo*-trimethylsilylbicyclobutonium ion **41** is visualized by NLMO calculations (CCSD/cc-pVTZ geometry) (Fig. 35A). The NBO occupation of the C1–C3 bridging bond in **41** is calculated to be 1.669 electrons and no bond is predicted for the C^3—Si bond but instead a lone pair at silicon which is calculated to contain only 0.340 electrons (Fig. 35B).

In the parlance of valence bond theory, this is depicted in the no-bond resonance structures **48a/48b**. This is in accord with the calculated NBO charges, which indicate that more than 60% of the positive charge residues in the trimethylsilyl substituent.

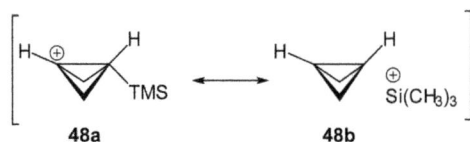

The calculated distance between the C_α and C_γ carbon in cation **41** (164.1 pm) is shorter than the C_α–C_γ distance calculated for the unsubstituted bicyclobutonium ion **9** (165.4 pm). This indicates a stronger bonding interaction between C_α and C_γ for **41**, which is due to the stronger stabilizing interaction of the *endo*-silyl group at the C_γ carbon with the formally positively charged carbon C_α as compared to the C_γ-*endo*–H–C_α interaction in **9**.

2.5 Spin–Spin Coupling Constants

Indirect carbon–carbon spin–spin coupling constants (SSCC) provide basic information about the nature of a chemical bond between carbon atoms, and thus the exploration of J_{CC} fundamentally would be another experimental tool to confirm the bridging bond in bicyclobutonium ions. Owing to the low probability of finding two ^{13}C carbon atoms next to each other (10^{-2}%) and to the experimental difficulties with low-temperature measurements on dilute solutions of reactive carbocations (which may not sustain long acquisition times), quantum–chemical calculations are a valuable alternative tool for the prediction of SSCCs of nuclei with low natural abundance.

J_{CC} values were calculated (CCSD/cc-pVTZ//ω97XD/pcJ-3) for model compounds such as cyclobutane **49**, bicyclo [1.1.0]butane **50**, the parent bicyclobutonium ion **9**, and the *endo*-3-trimethylsilyl-substituted bicyclobutonium ion **41**.

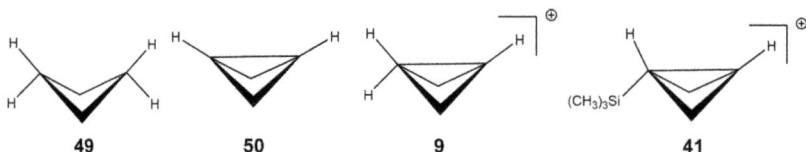

The $J_{(C1,C3)}$ coupling constants for the bicyclic structures **50** (-15.3 Hz), **9** (-12.7 Hz), and **41** (-15.2) are similar. The negative sign is characteristic for small-ring bicyclic molecules and is attributed to interference between one-bond and two-bond couplings in this type of small-ring system. The size of the J_{CC} coupling constant alone is thus not unequivocal proof for the bridging bond between C1 and C3, because of multiple pathways of the coupling interaction in small bicyclic ring systems. The coupling pathway can however be visualized by coupling deformation density (CDD) calculations. The Fermi-contact (FC) contribution to the J-coupling is related to the energy splitting between two states, with nuclear magnetic moments being antiparallel and parallel. The CDDs and the FC part of the one-bond coupling $^1J_{FC(C1-C3)}$ were calculated using a PERDEW exchange-correlation functional, and an IGLO basis using the deMon DFT code.[65] Visualization of the density difference provides a picture of the spin–spin interaction. The shape of the total CDD is dictated mainly by the contribution coming from the bond connecting the coupled nuclei.

Fig. 36 shows the CDD plots for cyclobutane **49**, a hypothetical planar cyclobutyl cation **8**, and bicyclobutane **50**, the parent bicyclobutonium ion **9**, and the *endo*-3-trimethylsilyl bicyclobutonium ion **41**.

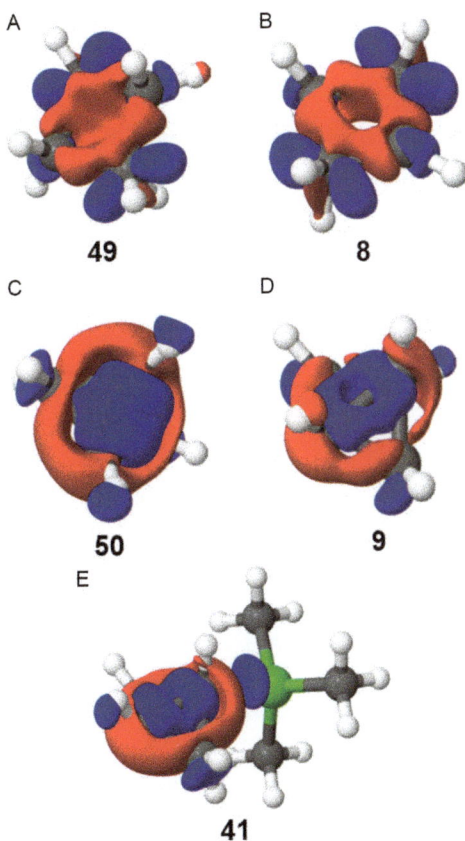

Fig. 36 (A) cyclobutane **49**, (B) hypothetical planar cyclobutyl cation **8**, (C) bicyclobutane **50**, (D) the parent bicyclobutonium ion **9**, and (E) the *endo*-3-trimethylsilyl bicyclobutonium ion **41**.

The $J_{(C1,C3)}$ of bicyclobutonium ions **9** and **41** confirm their bridged structure with a hypercoordinated C-4 carbon atom. This is supported by the similarity of the CDD plots for **9** and **41** with that of bicyclobutane **50** and the remarkable difference from the $J_{(C1,C3)}$ CDD plots of cyclobutane **49** and the planar cyclobutyl cation **8**.

In accord with the NBO calculation for the *endo*-3-trimethylsilyl bicyclobutonium ion **41**, i.e., the almost vacant orbital at silicon (Fig. 36) and the corresponding VB "no-bond" limiting structure **48b**, a rather small $^1J_{(Si,C)}$ SSCC of -9.3 Hz (CCSD/cc-pVTZ//ωB97XD/pcJ-3) is calculated for the Si—C3 bond in **41**.

2.6 1,3-Disubstituted Static Bicyclobutonium Ions

The γ-silyl effect, which leads to the static structure of 3-silyl substituted bicyclobutonium ions **41** and **43**, was used to generate other static bicyclobutonium ions, the 1-methyl,3-trimethylsilyl-1-bicyclo[1.1.0] butonium ion **50**[66] and the 1-trimethylmethyl,3-trimethylsilyl-1-bicyclo [1.1.0]butonium ion **51**.[67]

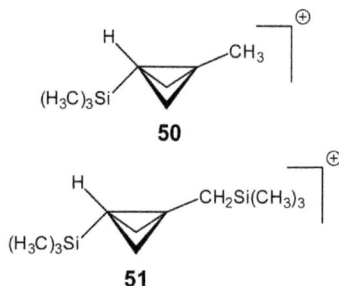

H. / $(H_3C)_3Si$ — CH₃ ⊕

50

H. / $(H_3C)_3Si$ — $CH_2Si(CH_3)_3$ ⊕

51

The ^{13}C NMR spectra of **50** (Fig. 37) and **51** (Fig. 38) clearly prove the static structure with the typical shielding of the pentacoordinated C_γ carbon.

A comparison of the ^{13}C NMR chemical shift of the pentacoordinated C_γ carbon in the bicyclobutonium ions **41**, **50**, and **51** ions, as well as a comparison of the ^{29}Si NMR chemical shifts and DFT calculations of the length of the bridging bonds C1–C3, show that with increasing electron-donating capability of the substituent at the C1 carbon from hydrogen to methyl and to a β-silyl group, the demand for γ-silyl stabilization of the positive charge is decreasing, which leads to a less deshielded ^{29}Si NMR signal for the γ-silyl substituent and a concomitant elongation of the C1–C3 bridging bond (Table 1).

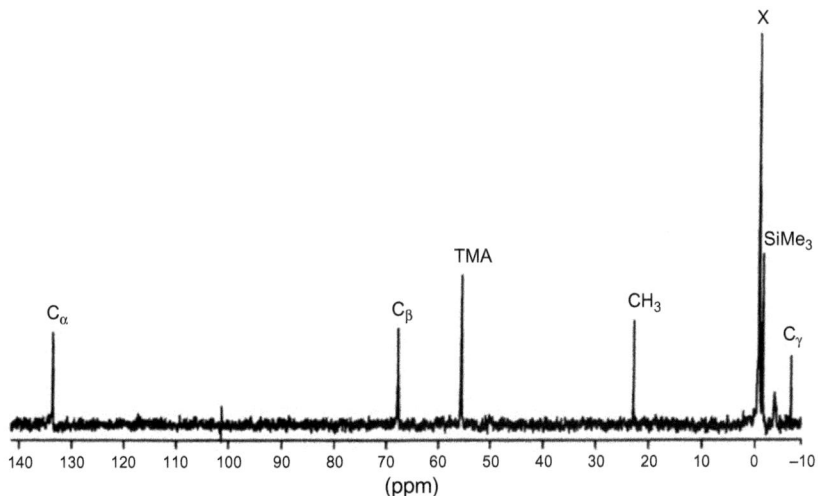

Fig. 37 ^{13}C NMR of 1-methyl,3-trimethylsilyl-1-bicyclo[1.1.0]butonium ion **50** (x=impurity).

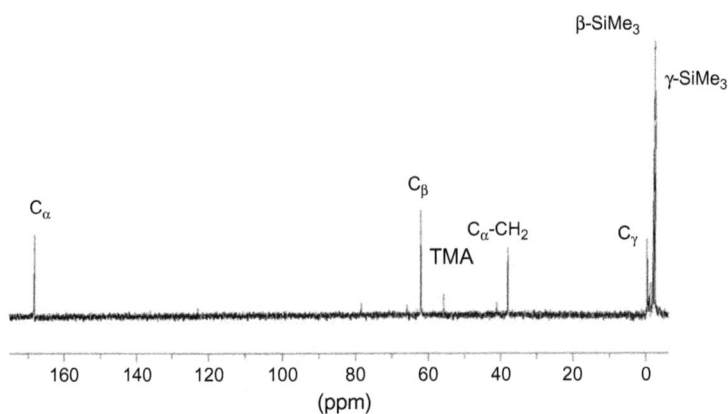

Fig. 38 ^{13}C NMR of 1-methyl,3-trimethylsilylmethyl-1-bicyclo[1.1.0]butonium ion **51**.

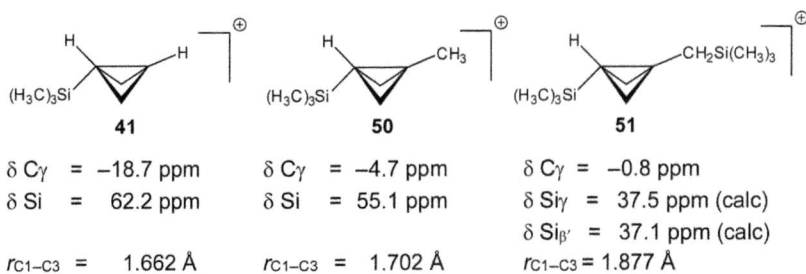

41	**50**	**51**
δ Cγ = −18.7 ppm	δ Cγ = −4.7 ppm	δ Cγ = −0.8 ppm
δ Si = 62.2 ppm	δ Si = 55.1 ppm	δ Siγ = 37.5 ppm (calc)
		δ Siβ′ = 37.1 ppm (calc)
$r_{C1–C3}$ = 1.662 Å	$r_{C1–C3}$ = 1.702 Å	$r_{C1–C3}$ = 1.877 Å

Table 1 Basic Properties of $C_4H_{7-n}R_n{}^+$ Bicyclobutonium Cations

Bicyclobutonium Ions Which Undergo Threefold-Degenerate Equilibration of CH_2-Groups	Static Bicyclobutonium Ions
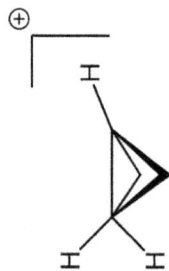 *No ring inversion*, stereospecific rearrangement, minor structural isomer cyclopropyl methyl cation **9**	 **41**
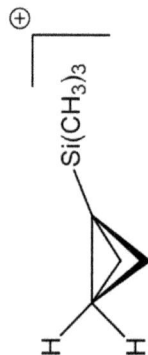 *No ring inversion*, stereospecific rearrangement, *no minor* isomer **33**	 **50**
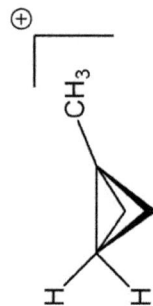 *Ring inversion no minor* isomer, static at ~ −153°C **28**	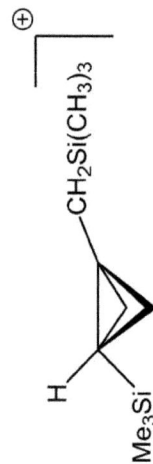 **51**

3. CONCLUSION

For over half a century, the area of physical organic chemistry of carbocations has triggered the development for better tools and methods. Exploring the chemistry of carbocations by NMR spectroscopy in superacid solutions in combination with today's state-of-the-art quantum-chemical calculations proved to be very successful and serves as a shining example for close integration of experimental and computational methods in physical organic chemistry. The investigation of EIEs on NMR spectra of fast-rearranging systems [32,33,35] has been of particular importance to solve the conundrum of $(C_4H_7)^+$ and related carbocations. The chemistry of hyper-coordinated carbocations, such as bicyclobutonium ions discussed in this chapter, has served as a forerunner for close integration of experimental and computational approaches applicable to all areas of chemistry.

Epilogue

Contrary to the alleged simplicity of low molecular weight structures such as $(C_4H_7)^+$ and related bicyclobutonium ions as reviewed in this chapter, it has been recognized that conventional bond formulas with single, double, and triple bonds are inadequate to account for the structure, dynamics, and properties of these (and many other) electron-deficient compounds. Hypercoordination of carbon, hydrogen, and other elements is no longer regarded as an exotic exception. As stated by G.A. Olah in his Nobel lecture "The long-time used prefixes '*classical*' and '*nonclassical*' are expected, however, to gradually fade away as the general nature of bonding will be recognized."[68]

REFERENCES

1. Surya Prakash GK, Schleyer PVR, eds. My personal relation to George Olah last 44 years and goes back to 1974 and details are published. *Stable Carbocation Chemistry*. New York: Wiley; 1997:165–166.
2. Bartlett PD. *Nonclassical Ions*. New York: W.A. Benjamin; 1965: 272.
3. Selected references:Wiberg KB, Hess Jr BA, Ashe AJ. Olah GA, Schleyer PVR, eds. *Carbonium Ions*. New York: Wiley; 1972:1295, vol 3. Lenoir D, Siehl, H-U. Houben-Weyl Methoden der Organischen Chemie. In: Hanack M, vol E19c, Stuttgart: Thieme; 1990: 413. Olah GA, Prakash Reddy V, Surya Prakash GK. Chem Rev 1992;92:69; Olah GA, Prakash GKS, Sommer J. Superacids. New York: Wiley; 1985: 143; Roberts JD. Proc Robert A Welch Found Conf Chem Res. 1990;34:312.
4. Brown HC. Schleyer PVR, ed. *The Nonclassical Ion Problem*. New York: Springer; 1977. with comments by.
5. Brown HC. The 2-norbornyl cation revisited. *Acc Chem Res*. 1986;19:34.
6. Roberts JD, Mazur RH. *J Am Chem Soc*. 1951;73:2509.

7. Olah GA, Juell CL, Kelly DP, Porter RD. *J Am Chem Soc*. 1972;94:146.
8. Roberts JD, Mazur RH. *J Am Chem Soc*. 1951;73:3542.
9. Mazur RH, White WN, Semenov DA, Lee CC, Silver MS, Roberts JD. *J Am Chem Soc*. 1959;81(1):4390–4398.
10. Trindle C, Sinanoglu O. *J Am Chem Soc*. 1969;91:4054.
11. Roberts JD. *The Right Place at the Right Time*. Washington, DC: American Chemical Society; 1990: 82.
12. Winstein S, Trifan D. *J Amer Chem Soc*. 1952;74:1145–1160.
13. Walling C. An innocent bystander looks at the 2-norbornyl cation. *Acc Chem Res*. 1983; 16:448–454. Brown HC. The 2-norbornyl cation revisited. Acc Chem Res 1986;19:34; Schleyer Paul von R, Mainz Vera V, Thomas Strom E. Norbornyl cation isomers still fascinate. In: The Foundations of Physical Organic Chemistry: Fifty Years of the James Flack Norris Award, Chapter 7: 139–168 ACS Symposium Series, vol 1209, American Chemical Society, 2015.
14. (a) Scholz F, Himmel D, Heinemann FW, Schleyer PvR, Meyer K, Krossing I. Crystal structure determination of the nonclassical 2-norbornyl cation. *Science*. 2013;341:62–64. (b) See however a recent view *J Phys Org Chem*. 2018; Hans-Jörg Schneider (accepted: 30 January 2018), https://doi.org/10.1002/poc.3846.
15. Joseph Casanova DR, Kent IV, William A, Goddard III, Roberts JD. *PNAS*. 2003;100(1):15–19.
16. Roberts JD. *J Organomet Chem*. 2009;74:4897–4917.
17. Minkin VI, Minkin RM, Minyaev Y, Zhdanov A. *Nonclassical Structures of Organic Compounds*. Moscow: Mir Publishers; 1987.
18. Goh SS, Champagne PA, Guduguntla S, et al. *J Am Chem Soc*. 2018;140:4986–4990.
19. Hong YJ, Giner J-L, Tantillo DJ. *J Am Chem Soc*. 2015;137:2085–2088.
20. McKee WC, Agarwal J, Schaefer III HF, Schleyer PvR. *Angew Chem Int Ed*. 2014;53:7875–7878.
21. Majerski Z, borcic S, Sunko DE. *Tetrahedron*. 1969;25:301–313.
22. Melander L, Saunders Jr WH. *Reaction Rates of Isotopic Molecules*. New York: Wiley; 1980. Collins CJ, Bowman NS, Eds. Isotope Effects in Chemical Reactions; ACS Monograph No. 167; Van Nostrand Reinhold: New York, 1970. Isotope Effects: in the Chemical, Geological, and Bio Sciences 2009th Edition by Max Wolfsberg, W. Alexander Van Hook (Author), Piotr Paneth (Author), Luís Paulo N. Rebelo (Author) Isotope Effects In Chemistry and Biology Nov 1, 2005 by Amnon Kohen and Hans-Heinrich Limbach.
23. Lit Goricnik B, Majerski Z, Borcic S, Sunko DE. *J Organomet Chem*. 1973;38:1881.
24. Bowen DHW, Schwarz H, Wesdemiotis C. *J Chem Soc Chem Commun*. 1979;6:261.
25. Holman RW, Pl°Cica J, Blair L, Giblin D, Gross ML. *J Phys Org Chem*. 2001;14:17–24.
26. Cacace F, Chiavarino B, Crestoni ME. *Chem Eur J*. 2000;6:2024–2030.
27. Vancik H, Gabelica V, Sunko DE, Buzek P, Schleyer PVR. *J Phys Org Chem*. 1993;6:427.
28. (a) Staral JS, Yavari I, Roberts JD, Prakash GKS, Donovan DJ, Olah GA. *J Am Chem Soc*. 1978;100(0):8016. (b) See however: Werstiuk NH, Poulsen DA. *ARKIVOC*. 2009;2008(v):38–50.
29. Myhre PC, Webb G, Yannoni CS. *J Am Chem Soc*. 1990;112:8992.
30. Martin Holzschuh PhD thesis, University of Ulm, 2016, H.-U. Siehl, M. Holzschuh, submitted for publication.
31. Lit:Jensen FR, Smith LA. *J Am Chem Soc*. 1964;86:956. Gold, 1968 (1968). Trans Faraday Soc 64, 2770.
32. Saunders M. Sarma R, ed. *Stereodynamics of Molecular Systems*. Oxford: Pergamon Press; 1979:171.

33. Carbocations, Fast rearrangements and the isotopic perturbation method M. Saunders O. Kronja chap. 8 213 in "Carbocation Chemistry" ed G.A. Olah G.K.S. Prakash, Wiley Interscience 2004.
34. Bogle S, Singleton DA. *J Am Chem Soc.* 2011;133:17172–17175.
35. Siehl H-U. *Adv Phys Org Chem.* 1987;23:63–163. and references therein.
36. Brittain WJ, Squillacote RJD. *J Am Chem Soc.* 1984;106:7280–7282.
37. Saunders M, Siehl H-U. *J Am Chem Soc.* 1980;102:6868–6869.
38. Saunders M, Laidig KE, Max Wolfsberg J. *Am Chem Soc.* 1989;111:8989–8994.
39. H.-U. Siehl, Habilitationsschrift Universität Tübingen, 1986.
40. Partial summary in:Surya Prakash GK. *J Organomet Chem.* 2006;71:3661–3676.
41. George A. Olah, G. K. Surya Prakash, and Golam Rasul J Am Chem Soc 130, 28, 9168–9172.
42. Stanton JF, Hans-UllrichSiehl JG. *Chem Phys Lett.* 1996;262:183–186.
43. Nikoletic M, Borcic S, Sunko DE. *Tetrahedron.* 1967;23(2):649–660.
44. Wiberg KB, Shobe D, Nelson GL. *J Am Chem Soc.* 1993;115:10645–10652.
45. Silver MS, Caserio MC, Rice HE, Roberts JD. *J Am Chem Soc.* 1961;83:3671.
46. M. Fuss, Ph.D. Thesis, University of Tübingen, 1997.
47. Saunders M, Rosenfeld J. *J Am Chem Soc.* 1970;92:2548.
48. Olah GA, Spear RJ, Hiberty PC, Hehre WJ. *J Am Chem Soc.* 1976;98:7470.
49. Olah GA, Surya Prakash GK, Donavan DJ, Yavari I. *J Am Chem Soc.* 1978;100:7085.
50. Kirchen RP, Sorensen TS. *J Am Chem Soc.* 1977;99:6687.
51. Saunders M, Norbert Kraus et J. *Am Chem Soc.* 1988;110:8050–8052.
52. Siehl HU. *J Am Chem Soc.* 1985;107:3390–3392.
53. Siehl H-U. Kobayashi M, ed. *Studies in Organic Chemistry.* Amsterdam: Elsevier; 1987:25–32. vol 31.
54. H.-U. Siehl Personal communication, 2018, unpublished.
55. Kostenko A, Müller B, Kaufmann F-P, Apeloig Y, Siehl H-U. *Eur J Org Chem.* 2012;9:1730–1736.
56. Siehl H-U, Fuss M, Gauss J, Am J. *Chem Soc.* 1995;117:5983–5991.
57. Josef Schneider 1985 Diplomarbeit Tübingen.
58. Siehl H-U, Koch E-W. *J Chem Soc Chem Commun.* 1985;8:496–498.
59. Koch E-W. *Dissertation.* Tübingen, Germany: University of Tübingen; 1985.
60. G. A. Olah, G. K. S. Prakash, and T. Nakajima, J Am Chem Soc,1 982, 104, 1031.
61. Schleyer PvR, Lenoir D, Mison P, Liang G, Prakash GKS, Olah GA. *J Am Chem Soc.* 1980;102:683.
62. Olah GA, Berrier AL, Arvanaghi M, Prakash GKS. *J Am Chem Soc.* 1981;103:1122.
63. Schneider J. *Dissertation.* University of Tübingen; 1989.
64. Siehl H-U, Fuss M. *Pure Appl Chem.* 1998;70:2015.
65. Malkina OL, Malkin VG. *Angew Chem Int Ed.* 2003;42:4335–4338.
66. Thomas Handel, Diplomarbeit University of Ulm, 2000.
67. Mato Knez Diplomarbeit University of Ulm, 2000.
68. Olah GA. *My Search for Carbocations and Their Role in Chemistry.* Nobel Lecture; December 8, 1994. https://www.nobelprize.org/prizes/chemistry/1994/olah/lecture/.

CHAPTER TWO

Organic Reaction Outcomes in Ionic Liquids

Rebecca R. Hawker, Jason B. Harper[1]

School of Chemistry, University of New South Wales, Sydney, NSW, Australia
[1]Corresponding author: e-mail address: j.harper@unsw.edu.au

Contents

Abstract

Ionic liquids have gained significant popularity in recent years as reaction media. Importantly, reaction outcomes (such as rates and selectivities) were shown to vary from traditional solvents. An understanding of such solvent effects in ionic liquids is required if these are to be used for preparative chemistry. This chapter discusses significant advances in this understanding that have led to ionic liquids being able to be used in a more predictive fashion along with a consideration of the opportunities that this new understanding may present into the future.

This work is dedicated to the late Ken Seddon, whose work in the area of ionic liquids was integral in establishing the field as it exists today. J.B.H. will never forget seeing Prof. Seddon speak at ICPOC15 (Göteborg, 2000) nor the conversations afterward that initiated the research programme discussed.

49

1. INTRODUCTION

Ionic liquids are arbitrarily defined as salts that have a melting point less than $100°C$[1]; they are sometimes referred to as room temperature molten salts.[2,a] The low melting points are a function of the structures of the constituent ions, with both the cations and anions frequently being large and charge-diffuse (and sometimes asymmetric).[3–7] Ionic liquids are not new; it can be argued that the first ionic liquid reported was in 1876 (picolinium citrate, by Ramsay[8]) though the first deliberate preparation (of ethylammonium nitrate) was in 1914 by Walden.[9] From these early "protic" ionic liquids, evolved Lewis acidic chloroaluminates (which suffered from water instability) and finally water-stable ionic liquids.[10,b] It is this last group, particularly their growing availability and the vast array of possible ion combinations,[11] which form the majority of the ionic liquids that have been considered extensively as solvents since the late 1990s.[c]

Ionic liquids might be thought of as potentially very attractive alternatives to molecular solvents.[12,13] They have extraordinarily low vapor pressures,[3,14–16] as is demonstrated in the extremes needed to distil them,[17] so are not readily lost to the environment (though there remain some toxicity concerns if they are released[18]) and are nonflammable, so safety concerns with generally flammable molecular solvents are ameliorated. Further their unique properties introduce the potential for recycling,[19,20] reducing concerns with their initial cost and removing issues associated with disposal of molecular solvents. As such, they are being considered in a range of roles, including as electrolytes[21] and catalysts,[22,23] along with acting as solvents for chemical processes.

When considered as solvents for preparative chemistry, it has been noted that reaction outcomes[d] in ionic liquids may differ significantly from those in molecular solvents.[24] An oft-quoted example is the nitration of toluene,

---

[a] This designation is perhaps somewhat misguided, as indicated by Welton: "After all, there is nothing special about room temperature, it just happens to be the temperature at which rooms are …."[2] And, across the world, rooms are rarely (if ever) at $25°C$.

[b] For a novice in the area of ionic liquids, this review is an excellent, personal account of the history of ionic liquids.

[c] The search term "ionic liquid" produces more than 80,000 results (in the years 2000–18) on ISI Web of Science.

[d] Examples of reaction outcomes reported include extents of reaction in a given time, rates of reaction, and positions of equilibria, with the former two being much, much more prevalent. Where isolated yields are reported, they must be carefully analyzed as any changes relative to a molecular solvent are a combination of solvent effects and ease of isolation.

where not only the rate and selectivity of nitration vary with the ionic liquid used but, in the extreme case, new reactivity (oxidation to give benzoic acid) is seen.[25] A synthetic chemist is unlikely to move to a solvent system where the outcome of reaction is not readily predicted, so a lack of understanding of solvent effects in ionic liquids can be considered a notable limitation to their application.

1.1 Scope of This Review

There are several, markedly different ways to consider solvent effects in ionic liquids. Reports observing changes in the isolated yield for a given process when carried out in an ionic liquid were prevalent in the early stages of ionic liquid research.[2] While these data are interesting in and of themselves, the lack of any control experiments (particularly on product isolation) and little (or no) discussion on the origin of the effects limits the utility of these reports. Studies in which ionic liquids are demonstrated to increase the rate and/or selectivity of a given reaction are more useful, though again the extent to which these contribute to the understanding of ionic liquid solvent effects is limited by the extent to which analyses of these effects are carried out. Of most use are those studies that attempt to determine the underlying molecular basis for how ionic liquids affect organic reactions, including considering microscopic interactions and solvent descriptors to explain ionic liquid solvent effects.

This review will cover the development of this understanding of ionic liquid solvent effects on reaction outcomes of organic processes. It is not intended to be comprehensive; there have been several notable reviews on ionic liquids, particularly their uses as solvents for synthetic procedures,[2,13,26,27] and our group has contributed several reviews on ionic liquid solvent effects[12,24,28–31]; readers are directed to those works if complete coverage of the literature is their aim. Rather, this work will outline how, through the building up of systematic data describing the effect of ionic liquids on well-understood reaction types, key features have been identified that can be used to predict (and exploit) the behavior of ionic liquids as solvents.

The range of effects discussed will be limited to those in which the solvent is not directly involved in the reaction. For an acid catalyzed process, it is not at all surprising that adding an acid to the reaction mixture would increase the rate of reaction—it is irrelevant whether that acid were sulfuric acid or an imidazolium hydrochloride. Such cases (where the solvents are often referred to as "task-specific ionic liquids"[32]) are unrelated to solvent effects of an ionic liquid and will not be covered here. Related systems like deep eutectic solvents[33] are also not discussed. Solvate ionic liquids, where one component is a complex ion typically formed by solvation of a metal ion such as lithium by a glyme,[34] might in future be included but there are currently limited reports of their use in preparative chemistry.[35–37]

1.2 Ionic Liquids: Common Structures and Nomenclature

As already introduced, charge-diffuse cations predominate in ionic liquids. Cations tend to have the greatest variation in structure, and they are almost exclusively organic in nature with a charged heteroatom (nitrogen and phosphorus being the most common). Some typical cations are shown in Fig. 1, along with abbreviations that are used for those species. Of these cations, the 1-alkyl-3-methylimidazolium series is arguably the most common and members of this series are frequently abbreviated [xmim]$^+$, where x is the first letter of the name of the alkyl chain. For example, the extremely widely used cation 1-butyl-3-methylimidazolium is abbreviated to [bmim]$^+$. Other, potentially less ambiguous,[e] abbreviations are also used; however, the [xmim]$^+$ nomenclature is common and will be used from this point on. Abbreviations for other systems, such as the pyrrolidinium cations, are determined in the same way; 1-butyl-1-methylpyrrolidinium is abbreviated to [bmpyr]$^+$. Abbreviations will be introduced as needed throughout this work.

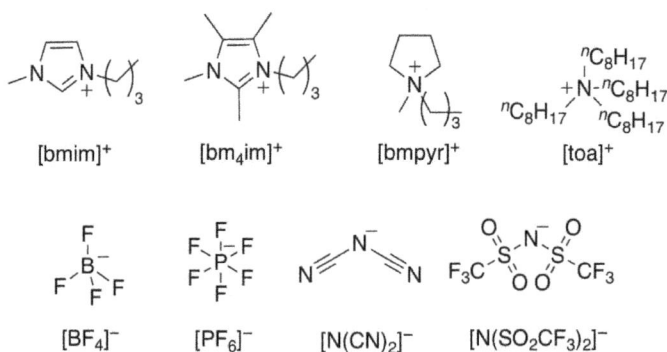

| [bmim]$^+$ | [bm$_4$im]$^+$ | [bmpyr]$^+$ | [toa]$^+$ |

| [BF$_4$]$^-$ | [PF$_6$]$^-$ | [N(CN)$_2$]$^-$ | [N(SO$_2$CF$_3$)$_2$]$^-$ |

Fig. 1 Examples of cations and anions used as components of ionic liquids.[1,10]

---

[e] For example, the imidazolium series can be abbreviated in the form [C$_X$C$_Y$im]$^+$, where X and Y are the number of carbons in the alkyl chains on the heterocyclic core. For example, 1-butyl-3-ethylimidazolium is abbreviated to [C$_4$C$_2$im]$^+$.

In contrast to the cations, anions are frequently referred to using their molecular formulas. The exceptions are some of the larger formula weight cations, where abbreviations are seen (such as TFSI for the very common bis(trifluoromethanesulfonyl)imide anion). As it is often useful to be able to immediately see the structure of the anion, the molecular formula approach will be used here.

2. UNDERSTANDING REACTION OUTCOMES IN IONIC LIQUIDS: KEY MILESTONES

A great deal of very careful work has been carried out in the last two decades in rationalizing reaction outcomes in ionic liquids. Among these studies, there have been several key points that are worth highlighting.

2.1 Not All Ionic Liquids Are Created Equal

The assumption that all ionic liquids affect reaction outcomes in the same fashion was demonstrated to be at best a poor predictor very early on in the study of ionic liquids as solvents for preparative chemistry. Many examples were present in the early ionic liquid literature demonstrating that different ionic liquids affected reaction outcome in different ways; some representative (and useful) examples are introduced here.

Diels–Alder processes were among the earliest reactions studied in ionic liquids and typically multiple ionic liquids were considered. A demonstrative reaction is that between cyclopentadiene 1 and methyl acrylate 2 (Scheme 1)[38–41]; other dienophiles have also been considered. For the reaction given, ionic liquids were shown to increase the rate of the reaction (as determined from isolated yields and measured rate constants) relative to molecular solvents and to vary the ratio of the isomers 3 and 4 formed. Importantly the rate enhancement and the selectivity varied with the ionic liquid considered.

Perhaps the example that is cited most often (possibly due to the somewhat declarative title of the work) is the nitration of toluene 5 reported by Seddon (Scheme 2).[25] This nitration process was found to proceed to give a regioisomeric mixture of nitrotoluenes 6, though in the presence of each of several ionic liquids (including [bmim][CF$_3$SO$_3$]) the rate of such formation (as measured by extents of conversion in a given time) was increased, as was the selectivity for the formation of the *ortho* isomer. Most notable from this study, however, was the observation that oxidation of toluene 5 to give benzoic acid 7 was observed in [bmim][CH$_3$SO$_3$];

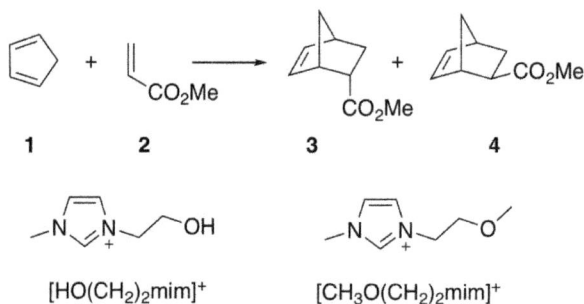

Solvent	Yield (%)	endo 3 : exo 4
Methanol[35]	n.r.	6.7 : 1
Ethanol[35]	n.r.	5.2 : 1
Acetone[35]	n.r.	4.2 : 1
Diethyl ether[35]	n.r.	2.9 : 1
[emim][CF$_3$SO$_3$][33]	56	4.9 : 1
[emim][PF$_6$][33]	34	3.2 : 1
[emim][NO$_3$][33]	57	3.3 : 1
[bmim][ClO$_4$][33]	75	5.3 : 1
[bmim][BF$_4$][33]	91	4.3 : 1
[HO(CH$_2$)$_2$mim][N(SO$_2$CF$_3$)$_2$][34]	n.r.	6.1 : 1
[CH$_3$O(CH$_2$)$_2$mim][N(SO$_2$CF$_3$)$_2$][34]	n.r.	5.7 : 1

Scheme 1 The Diels–Alder reaction between cyclopentadiene **1** and methyl acrylate **2** to give the stereoisomers **3** and **4**, the yields after 72 h at 25°C and the ratios of products **3** and **4** in the solvents shown (along with the structures of two-key cations). n.r.—not reported under the same conditions as in Ref. 33.

this was the first demonstration of a completely different product when an ionic liquid was added. Interestingly, no rationalization of this outcome was given.

2.2 The Importance of the Proportion of the Ionic Liquid in the Reaction Mixture

Many reports in the literature purport that a reaction is "carried out in an ionic liquid." Such a statement should be looked at very carefully (as should the appropriate Experimental section in these manuscripts) before continuing because, in many cases, the reaction mixture is notably more complicated than that simple description, with the ionic liquid significantly

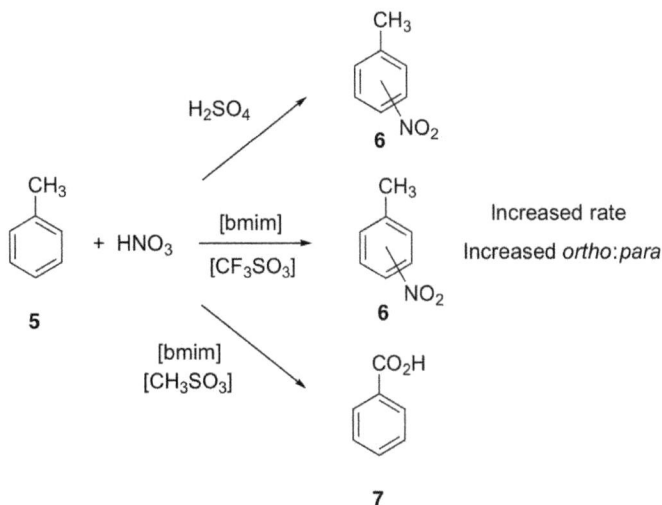

Scheme 2 The reaction of toluene **5** with nitric acid, which was found to proceed differently in different ionic liquids.[25]

diluted by either a cosolvent, reagents, or both. In such cases, the proportion of ionic liquid in the reaction mixture (particularly as a mole fraction, noting the large formula weight of most ionic liquids) may be small,[f] with the reaction mixture effectively becoming a dilute salt solution. It is immediately worth pointing out that this does not invalidate the results observed—the outcomes are what the outcomes are—but that an understanding of the effects of varying the proportion of the ionic liquid in the reaction mixture on reaction outcome is important. However, it should be considered whether the term ionic liquid is really how additives present at small concentrations should be promoted; typically the fact that the species is a salt that is soluble in an organic solvent is important, that the species is a liquid below 100°C is not.

The importance of the amount of ionic liquid in the reaction mixture was first identified in the unimolecular substitution process involving the linalool derivative **8** transformed to the ether **9** (Scheme 3). Importantly, the rate constant of the process was observed to vary with the proportion of the ionic liquid [bmim][N(SO$_2$CF$_3$)$_2$] in the reaction mixture (Fig. 2).[42] At low proportions of ionic liquid in the reaction mixture, a rate constant increase relative to the molecular solvent, methanol, was observed, while at high proportions of the ionic liquid in the reaction mixture a

---

[f] Typically, reaction mixtures have to be >80% ionic liquid by volume to be $\chi_{\text{ionic liquid}} > 0.5$.

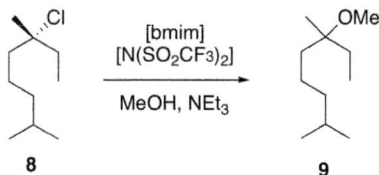

Scheme 3 A unimolecular substitution process that has been examined in mixtures containing the ionic liquid [bmim][N(SO$_2$CF$_3$)$_2$].[42]

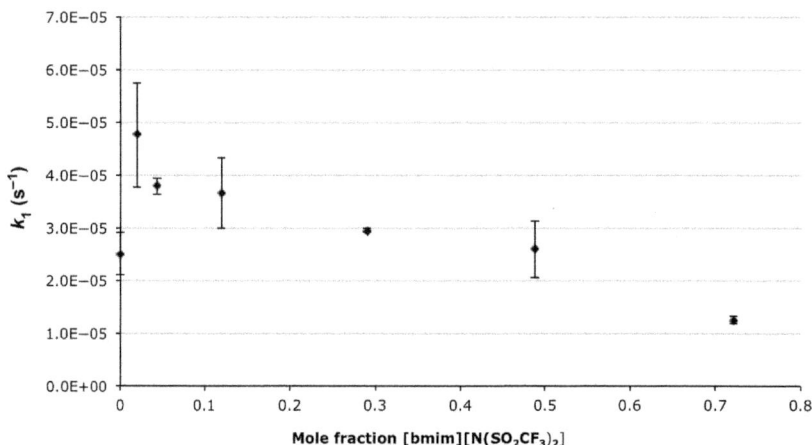

Fig. 2 The unimolecular rate constant for the reaction of the chloride **8** with methanol in different mole fractions of the ionic liquid [bmim][N(SO$_2$CF$_3$)$_2$] in methanol. *Reproduced from Man BYW, Hook JM, Harper JB. Substitution reactions in ionic liquids. A kinetic study. Tetrahedron Lett. 2005;46:7641–7645 with the permission of Elsevier.*

decrease in the rate constant was observed.[g] Significantly, this example demonstrates that knowing the proportion of ionic liquid in the reaction mixture is important; the ionic liquid could be considered to either enhance or decrease the rate constant depending on what proportion of ionic liquid is present.[43]

While this was the first demonstration of the importance of understanding the proportion of the amount of ionic liquid in the reaction mixture, subsequent studies have shown how reaction outcome changes for other processes including intramolecular nucleophilic displacement,[44] another unimolecular

---

[g] These data are consistent with the observed reduction in the unimolecular rate constant for the hydrolysis of 1-adamantyl mesylate in [bmim][N(CF$_3$SO$_2$)$_2$] ($\chi_{salt} > 0.8$) relative to common polar organic solvents.[43] Of interest, those authors speculated (without evidence) on the ionic liquid ordering about the transition state of the process (see Section 2.3).

substitution reaction,[45] bimolecular nucleophilic substitution,[45–48] nucleo-philic aromatic substitution,[49,50] and bimolecular condensation processes,[51–53] reactions at phosphorus centres,[54–56] and cycloaddition reactions.[12,57,h] Of note is that each of these different processes has a different dependence of reaction outcome on the proportion of the salt in the reaction mixture; two examples which have been considered extensively (and will be referred to subsequently) are the reaction of pyridine **10** and benzyl bromide **11** to give the salt **12** (Scheme 4, Fig. 3)[47] and the ethanolysis of the fluorobenzene **13**

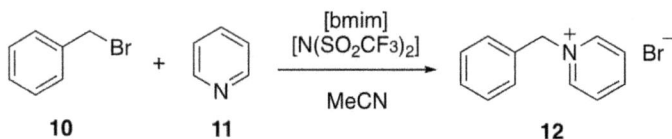

Scheme 4 A bimolecular substitution process that has been examined in mixtures containing the ionic liquid [bmim][N(SO$_2$CF$_3$)$_2$].[47]

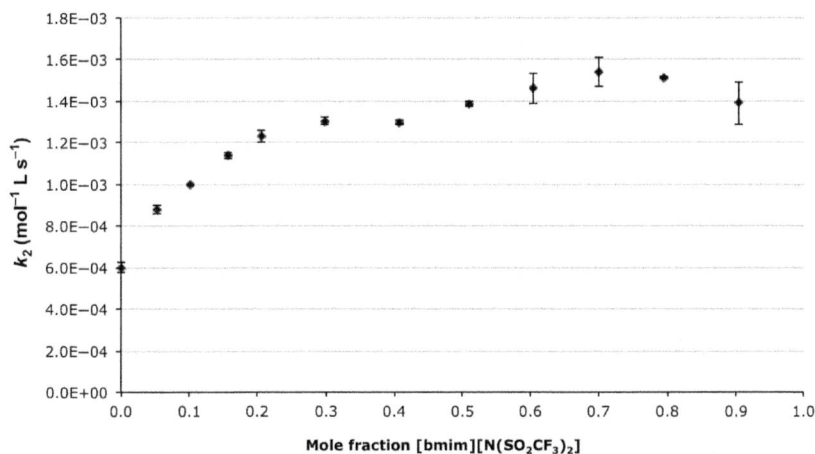

Fig. 3 The bimolecular rate constant for the reaction of the bromide **10** with pyridine **11** in different mole fractions of the ionic liquid [bmim][N(SO$_2$CF$_3$)$_2$] in acetonitrile. The equivalent trend for the corresponding chloride was also reported earlier.[46] *Reproduced from Schaffarczyk McHale KS, Hawker RR, Harper JB. Nitrogen versus phosphorus nucleophiles—how changing the nucleophilic heteroatom affects ionic liquid solvent effects in bimolecular nucleophilic substitution processes. New J Chem. 2016;40:7437–7444 with permission from the Centre National de la Recherche Scientifique (CNRS) and the Royal Society of Chemistry.*

---

[h] It should be noted that some of these reports use volume fraction rather than mole fraction, which complicates comparison between different cases.

to give the phenetole **14** (Scheme 5; Fig. 4).[49] Together, these three examples represent three different profiles of how reaction outcome may vary with the composition of the reaction mixture.

It should be noted that, given the differences in reaction outcome noted between different ionic liquids, all of the examples above used the same ionic liquid ([bmim][N(SO$_2$CF$_3$)$_2$]),[i] thus demonstrating that the changes are not

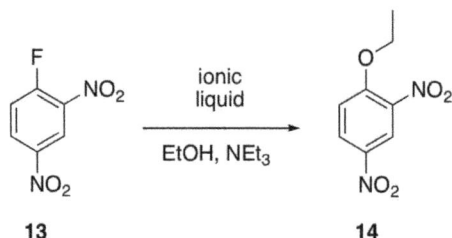

Scheme 5 A nucleophilic aromatic substitution process that has been examined in mixtures containing one of a range of ionic liquids.[49,50]

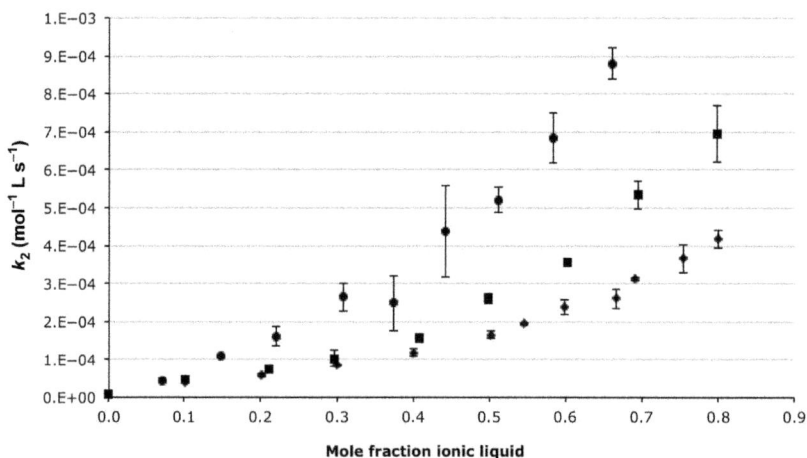

Fig. 4 The bimolecular rate constant for the reaction of the fluorobenzene **13** with ethanol in different mole fractions of either [bmim][N(SO$_2$CF$_3$)$_2$] (◆), [bm$_4$im][N(SO$_2$CF$_3$)$_2$], (■) or [toa][N(SO$_2$CF$_3$)$_2$] (●) in ethanol. *Adapted from Hawker RR, Wong MJ, Haines RS, Harper JB. Rationalising the effects of ionic liquids on a nucleophilic aromatic substitution reaction. Org Biomol Chem. 2017;15:6433–6440 with permission from the Royal Society of Chemistry.*

---

[i] This ionic liquid is extremely common in the literature as it is easy to prepare[58] (particularly in terms of removing impurities that can affect physical properties of,[59] and reaction outcomes in,[60] ionic liquids) and has excellent chemical and thermal stability.[61]

the result of changing the ionic liquid. Generally, the mole-fraction dependence of reaction outcome has been shown to be independent of the components of the ionic liquid; that is, while the magnitude of any changes in, say, rate constant may vary as the constituent ions are varied, the general trend of such plots is the same. This feature is illustrated in Fig. 4 for the substitution process shown in Scheme 5,[49] but has also been shown for bimolecular nucleophilic substitution reactions,[48] bimolecular condensation processes,[51–53] and reactions at phosphorus centres.[54–56,58–61]

2.3 Understanding the Microscopic Origins of Ionic Liquid Effects

The above two sections demonstrate key observations about ionic liquid solvent effects. However, in order to generate any sort of predictive framework for such effects they must be rationalized and a clear way of doing so is to consider the interactions occurring at the microscopic level.

Early efforts to identify microscopic interactions responsible for ionic liquid solvent effects used a combination of carefully designed experiments and chemical intuition. For example, the Diels–Alder reaction described in Scheme 1 was shown to be favored by ionic liquids containing coordinating cations (particularly through hydrogen-bond donation) and those containing extremely poorly coordinating anions; these salts also gave greater selectivity for the *endo* isomer **3**. Given that Lewis acids are known to increase the selectivity and rates of Diels–Alder reactions,[62] it was proposed that rate enhancement and increase in selectivity were due to interaction of the cation with the oxygen on the substrate **2**.[40] This proposal is consistent with the observation that this reaction type has also been found to be accelerated in chloroaluminate ionic liquids,[63–65] and is supported by decreases in the selectivity in the presence of added charge-dense ions.[41] However, enhanced selectivities have been seen in some ionic liquids with greatly reduced hydrogen-bond donor ability[41,66] and the same argument cannot apply for all cycloaddition processes as some that are slowed by the addition of Lewis acids are accelerated in ionic liquids (see, for example, nitrile oxide cycloaddition reactions[67]).

Also early on in the ionic liquid literature it was frequently proposed that ionic liquids would stabilize charges, and thus accelerate reactions that involved charge development.[68] While this understanding parallels what is frequently used in the literature for traditional molecular solvents[69] it would subsequently be shown as a simplification, with activation-parameter data (enthalpies and entropies of activation) determined through temperature-dependent kinetic analyses for a range of reactions.

It is important to recognize that changes in the activation parameters on moving from a molecular solvent to a reaction mixture containing an ionic liquid represent a change in the difference in either enthalpy or entropy between the starting materials and the transition-state complex. As such, for example, an increase in the enthalpy of activation could be the result of either a decrease in the enthalpy of the reagents,[j] an increase in the enthalpy of the transition state or some combination of both. Given that ionic liquids are made up of charged components, it is most likely that the introduction of such components to the reaction mixture would increase interactions with other species in solution, rather than decrease them. With this in mind, an increase in the activation enthalpy on moving from a molecular solvent to an ionic liquid suggests greater interaction of the components of the ionic liquid with the starting materials, unless other factors can be identified; the converse can be argued for a decrease in an activation parameter. Significantly, these data allow identification of the principal interaction responsible for the changes in rate constants observed.

Significant early work was done by the Welton group in determining the activation parameters for bimolecular nucleophilic substitution processes.[70] A demonstrative example is the reaction between the amine **15** and methyl nosylate **16** (Scheme 6); this reaction was found to proceed with a greater

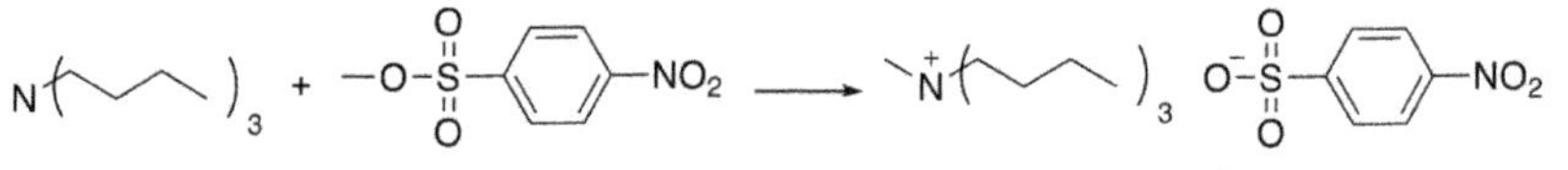

Solvent	$\Delta H^{\ddagger}$ (kJ mol^{-1})	$\Delta S^{\ddagger}$ (J K^{-1} mol^{-1})
Dichloromethane	34.5 ± 2.2	−161 ± 8
Acetonitrile	35.7 ± 0.8	−155 ± 4
[bmpyr][CF$_3$SO$_3$]	59.4 ± 2.0	−52 ± 7
[bmim][CF$_3$SO$_3$]	50.4 ± 4.3	−100 ± 14

Scheme 6 The bimolecular nucleophilic substitution reaction between tributylamine **15** and methyl *p*-nitrobenzenesulfonate **16** which has been studied in the ionic liquids [bmpyr][CF$_3$SO$_3$] and [bmim][CF$_3$SO$_3$], along with the activation parameters determined for the reaction in a range of solvents.[70]

---

[j] Strictly this could also be due to a greater decrease in the enthalpy of the reagents than the transition state but, as is used in the following argument, identifying the principle interaction is considered most important.

rate constant in reaction mixtures containing the ionic liquids [bmpyr] [CF$_3$SO$_3$] and [bmim][CF$_3$SO$_3$] (χ_{salt} ca. 0.80). Activation–parameter data showed that there were increases in both the enthalpy and entropy of activation compared to the molecular solvents considered. These data were consistent with greater interactions of the ionic liquid solvents with the starting materials than the molecular solvents; particularly, interactions of the cation with the nitrogen center of the nucleophile **15**. It is worth noting that the rate enhancement was due to an entropic effect; that is, not due to the stabilization of the developing charges in the transition state.[71–74,k]

Similar results were obtained for the bimolecular substitution reaction shown in Scheme 4; there was an increase in both activation parameters on addition of the ionic liquid [bmim][N(SO$_2$CF$_3$)$_2$] to the reaction mixture (χ_{salt} ca. 0.80).[75] In this case, it was not immediately apparent which starting material (if not both) was involved in the key interactions with the ionic liquid, though it should be noted that the interaction must be disrupted on reaction occurring. Given this as a starting point, it was unlikely that the key interaction was between the ionic liquid and the quadrupole of the aromatic portion of the benzyl bromide **10** and interaction with the pyridine **11** was thought most likely. Deconvolution studies involving variation of both reagents suggested the key interaction was between the cation of the ionic liquid and the lone pair on the pyridine **11** (cf. potential interaction with the delocalized π system on the pyridine **11**).[76] Such arguments were supported by molecular dynamics simulations, which provided visual representations of the interactions present in solution (Fig. 5).

These arguments regarding the microscopic origins of the solvent effects on this reaction were extended to consider the site of interaction on the ionic liquid cation. Through considering a range of solvents containing different cations (with different steric requirements and degrees of charge localization), a clear trend in the rate constant was seen.[77] The largest rate constant was seen with the pyrrolidinium derivative [bmpyr][N(SO$_2$CF$_3$)$_2$], with decreasing rate constants seen with imidazolium derivatives (with greater substitution resulting in smaller rate constants), while [toa][N(SO$_2$CF$_3$)$_2$] resulted in the smallest rate constant. While in all cases the rate constant enhancements relative to the molecular solvent were entropically driven, the (relatively) small changes in rate constant between the different ionic liquid

---

[k] The same group also examined extensively reactions of the same electrophile **16** with anionic nucleophiles; in those cases, significant concentrations of ionic liquids resulted in ion pairing which slowed the reaction relative to non–coordinating solvents. These interactions were borne out in the activation parameters.[71–74]

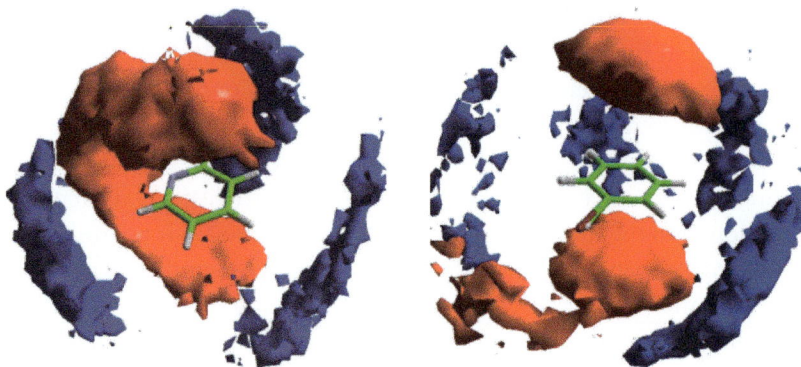

Fig. 5 Molecular dynamics simulations of the components of an ionic liquid showing the probability densities ([bmim]$^+$: *red*; [N(SO$_2$CF$_3$)$_2$]$^-$: *blue*; cutoff for both: 0.05) around pyridine **11** (*left*) and benzyl bromide **10** (*right*). *Images reproduced from Yau HM, Croft AK, Harper JB. Investigating the origin of entropy-derived rate accelerations in ionic liquids.* Faraday Disc. *2012;154:365–371 with permission from the Royal Society of Chemistry.*

cases and uncertainties in activation-parameter data made comparison of enthalpies and entropies of activation unenlightening. Irrespective, this outcome identified that the key site of interaction on the ionic liquid cation was the charged center.[78,1]

The above examples looked at activation parameters at a single composition of solvent; as demonstrated in the previous sections, it is important to consider how activation parameters might change with the proportion of ionic liquid in the reaction mixture. For the S_N1 reaction shown in Scheme 3, there was an initial rate-constant increase on addition of [bmim] [N(SO$_2$CF$_3$)$_2$] to the reaction mixture.[42] While no change in activation parameters between the methanol case and the χ_{salt} ca. 0.02 case were observed, at χ_{salt} ca. 0.50 there was a decrease in both the enthalpy and entropy of activation. It was argued that the ionic liquid [bmim][N(SO$_2$CF$_3$)$_2$] was interacting with, and organizing around, the transition state of the process.[79] This argument was considered reasonable given the significant charge development in the transition state; at low mole fractions of the salt, enthalpic contributions dominate resulting in a rate constant enhancement while at high proportions of the salt in the reaction mixture the organization of the components of the ionic liquid around the transition state dominates and

---

[1] No systematic variation in the rate constant was observed on varying the anion of the ionic liquid[78]; this demonstrated that changing the anion affects more than just the coordinating ability of the system and is likely due to the vastly differing sizes and shapes of the anions used (cf. the cations).

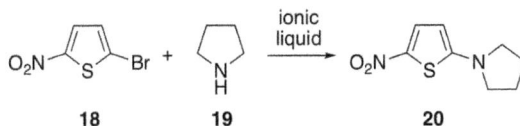

Scheme 7 A nucleophilic aromatic substitution process that has been examined in mixtures containing either [bmim][BF$_4$] or [bmim][PF$_6$].[80]

the decrease in the entropy of activation (an entropic cost) resulted in a rate-constant decrease. This case further disproves the argument that stabilization of charge development results in rate enhancement.

Similar studies have been carried out for nucleophilic aromatic substitution processes. Decreases in the enthalpy and entropy of activation, relative to the methanol case, for the reaction of the thiophene **18** with pyrrolidine **19** to give the substituted system **20** (Scheme 7) were seen on addition of ionic liquids [bmim][BF$_4$] and [bmim][PF$_6$] (χ_{salt} ca. 0.75 in both cases).[80] While these data are useful and suggest interaction with, and stabilization of, the charge-separated transition state (resulting in an enthalpy-based rate enhancement), this was carried out at only one reaction mixture composition containing each ionic liquid.

Of interest are the studies that focused on the change in activation parameters at different compositions of solution for the reaction shown in Scheme 5. When the reaction was carried out in the ionic liquid [bmim][N(SO$_2$CF$_3$)$_2$] at χ_{salt} ca. 0.50, both the enthalpy and entropy of activation increased relative to the molecular solvent case.[81] Consistent with the key interaction involving the ionic solvent being with starting materials, it was proposed that the changes in activation were due to organization of the solvent around the benzene **13**,[m] with the entropy gained on disrupting this organization responsible for the observed rate-constant enhancement. This argument was supported by molecular dynamics simulations of the salt about both the starting material **13** and the Meisenheimer intermediate.[n]

Subsequent studies considered the activation parameters at different mole fractions of each of a range of ionic liquids (representative data are shown in Table 1).[49,50,77,81] Several things should be noted from these data: (i) the changes in activation parameters are not consistent over the range of solvent compositions studied, with different changes occurring at different solvent compositions; (ii) the nature of the ionic liquid affects the extent to

---

[m] From the data, it was not clear which of components of the ionic liquid (or both) were important in the interaction.

[n] The intermediate served as a surrogate for the transition state.

Table 1 The Activation Parameters for the Reaction Between the Fluorobenzene **13** and Ethanol in Either Ethanol or in the Ionic Liquid Mixture Specified

Solvent	χ_{salt}	$\Delta H^{\ddagger}$/kJ mol^{-1}	$\Delta S^{\ddagger}$/J K^{-1} mol^{-1}
Ethanol[a]	0	49.0 ± 0.5	−259.0 ± 1.6
[bmpyr][N(SO$_2$CF$_3$)$_2$][b]	0.51	45.7 ± 1.9	−240 ± 6
[bmpyr][N(SO$_2$CF$_3$)$_2$][a]	0.78	43.0 ± 2.3	−242 ± 7
[bmim][N(SO$_2$CF$_3$)$_2$][c]	0.54	49.6 ± 0.5	−229 ± 2
[bmim][N(SO$_2$CF$_3$)$_2$][a]	0.79	44.6 ± 1.8	−238 ± 6
[bmim][N(CN)$_2$][b]	0.63	58.0 ± 3.1	−201 ± 10
[bmim][N(CN)$_2$][d]	0.80	41.7 ± 2.7	−246 ± 8
[toa][N(SO$_2$CF$_3$)$_2$][b]	0.32	41.0 ± 2.6	−252 ± 8
[toa][N(SO$_2$CF$_3$)$_2$][a]	0.66	32.4 ± 1.0	−270 ± 3

[a]Data from Hawker et al.[49]
[b]Data from Tanner et al.[77]
[c]Data from Jones et al.[81]
[d]Data from Hawker et al.[50]

Table 2 The Activation Parameters Responsible for the Rate-Constant Enhancement for the Reaction Between the Fluorobenzene **13** and Ethanol in Each of the Ionic Liquids Shown for the Given Change in Mole Fraction[49,50]

Solvent	χ_{salt} 0 → 0.5	χ_{salt} 0.5 → 0.8	χ_{salt} 0 → 0.8
[bmpyr][N(SO$_2$CF$_3$)$_2$]	$\Delta H^{\ddagger}$ and $\Delta S^{\ddagger}$	—[a]	$\Delta H^{\ddagger}$ and $\Delta S^{\ddagger}$
[bmim][N(SO$_2$CF$_3$)$_2$]	$\Delta S^{\ddagger}$	$\Delta H^{\ddagger}$	$\Delta H^{\ddagger}$ and $\Delta S^{\ddagger}$
[bmim][N(CN)$_2$]	$\Delta S^{\ddagger}$	$\Delta H^{\ddagger}$	$\Delta H^{\ddagger}$ and $\Delta S^{\ddagger}$
[toa][N(SO$_2$CF$_3$)$_2$]	$\Delta H^{\ddagger}$	$\Delta H^{\ddagger}$	$\Delta H^{\ddagger}$

[a]No change in activation parameters was seen in this range.

which activation enthalpy and activation entropy contribute to any observed change; and (iii) while rate-constant enhancements relative to molecular solvent are seen across all mole fractions of all of the ionic liquids considered, whether these are enthalpy or entropy driven depends on both the nature of the salt added and its proportion in the reaction mixture.[49,50]

This last point is illustrated well in Table 2, which shows the activation parameter(s) responsible for the rate-constant enhancement observed on changing the solvent composition. Importantly, these data demonstrate that both interactions with the starting material (which result in an entropic benefit) and interactions with the transition state (which result in an enthalpic benefit) are contributing to the observed effects. It should be noted that the sites of interaction on the ionic liquid are not explicitly determined but the (significant) differences observed on changing either ionic component of the ionic liquid suggests that both components are involved.

It is worth reflecting briefly on all of the examples the above section; given the magnitude of rate constant changes that are observed on addition of an ionic liquid to the reaction, the difference in the Gibbs energy of activation between two solvent compositions is comparatively small. Bearing this in mind, along with uncertainties in activation-parameter data, it is notable that differences in activation enthalpy and entropy are observed in these cases.° This is due to the fact that the interactions involving the ionic liquids are significant and result in significant stabilization of, and ordering about, species along the reaction coordinate.[60,67,82]

To conclude, it is worth considering a final point. The arguments presented do not rely on the physical state of the salt being used; the fact that the examples given are ionic liquids is beside the point.[83] In fact, in many of the cases presented, "normal" salts (with high melting points) were considered as additives to the reaction mixture, and in some cases (such as the nucleophilic aromatic substitution reaction[50]) they resulted in rate-constant changes comparable or greater than ionic liquids at equivalent solvent compositions. That last phrase, however, highlights one of the key advantages of ionic liquids; as they are liquid, they can be present in solution at a much higher proportion than salts that are solid at room temperature, thus increasing their potential to affect reaction outcome.

2.4 Correlating Reaction Outcomes With Solvent Parameters

Molecular solvents are often characterized by solvent parameters, which are quantitative measures of the capability of solvents for interaction with solutes.[84] It would clearly be useful if reaction outcomes in ionic liquids could be correlated to such parameters because an appropriate ionic liquid could be chosen to get the desired outcome.

One term that is often used when considering solvents is polarity. Polarity is defined as "[a measure of] the nonspecific attractive and repulsive forces, essentially of an electrostatic nature, between ionic or dipolar

---

° Enthalpies and entropies of activation do not always differ, even when rate-constant changes are seen; for examples see Refs. 60, 82. In these cases, and others where determining activation parameters is not practical,[67] assessing the microscopic origin of ionic liquid solvent effects is complicated.

solutes and solvent dipoles".[85] Various solvent parameters (such as the Dimroth–Reichardt E_T parameter) might be considered to measure polarity, but the "apparent polarity" is clearly very dependent on the portion of the solvent interacting with the probe molecule[86] and modeling such parameters is problematic at best.[87] When such measurements are carried out (for examples, see Welton[88] and references cited therein), the values are generally found to be comparable to short chain alcohols. Unfortunately, frequently the reaction outcomes seen in ionic liquids (particularly changes with composition of the reaction mixture but also the microscopic origins) do not correlate well with polarity measurements (see, for example, the case of the unimolecular substitution reaction seen in Scheme 3[42,79]).

Perhaps unsurprisingly given the work described in this review, more success has been had with parameters that measure (or at least contend to measure) interactions in solution that might be responsible for the changes in reaction outcomes. One set of solvent parameters that have been considered repeatedly[89] are the Kamlet–Taft solvent parameters[90–92]; α (a measure of the hydrogen-bond donating ability of the solvent), β (a measure of the hydrogen-bond accepting ability of the solvent), and π^* (a measure of the overall polarizability of the solvent). The Diels–Alder reaction introduced earlier (Scheme 1), along with equivalent reactions using related dienophiles, has been examined in an effort to develop linear solvation energy relations using these parameters.[93] Given the arguments presented on the proposed origin of the solvent effects, it is perhaps not surprising that the selectivity could be correlated relatively well to a linear combination of the parameters with the most significant contribution coming from the α parameter. Correlations involving the rate constants for the processes were far less clear, and there were significant differences in weighting of parameters between different substrates, clearly highlighting the limitations of the model.[P] Interestingly, subsequent computational modeling suggested that other interactions also contribute to the observed effects[95] and the dependence on the α parameter may be related to preferential interaction between components of the solvent.[96] Other examples of correlations with Kamlet–Taft parameters are also included where appropriate in subsequent sections.[94]

[P] Interestingly, when an intramolecular case was considered, correlations simply with Kamlet–Taft parameters were fair; inclusion of a viscosity term improved the correlations dramatically but the resulting fit showed limited dependence on the α parameter, raising the question of correlation vs causation.[94]

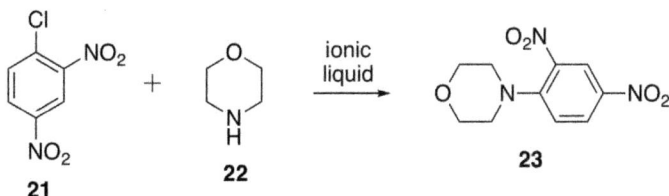

Scheme 8 A nucleophilic aromatic substitution process between the activated benzene **21** and morpholine **22** to give the aniline **23**, which been examined in an extensive range of ionic liquids.[98]

An alternative measure that might be used in similar situations where the Lewis basicity is important is the donor number of the anion (sometimes referred to as the Gutmann donor number, named after the discloser of the method[97]). The nucleophilic aromatic substitution shown in Scheme 8 is an example of a reaction where the anion of the ionic liquid was shown to be more significant in affecting reacting outcome than the cation.[98] The rate constant for the reaction in ionic liquids was found to correlate well with the donor number of the anion of the ionic liquid.[q] Subsequently it was demonstrated that rate constants for a related nucleophilic aromatic substitution process discussed earlier (Scheme 5) can also be reasonably correlated to the Kamlet–Taft β parameter.[50]

3. OPPORTUNITIES DELIVERED BY THE UNDERSTANDING GAINED

Given the increase in understanding gathered it is useful to consider how this might be exploited to use ionic liquids to control reaction outcome. Potential opportunities have been grouped below, based on the level of generality but also the extent to which they have already been developed.

3.1 Extending the Knowledge to Other Systems

With the understanding of the microscopic origins of ionic liquid solvent effects developed above, it is now possible to make (albeit somewhat general) statements about how an ionic liquid might affect the reaction outcome of a given process. For example, with the understanding shown for unimolecular substitution processes,[42,79] it might be reasonable to assume

---

[q] There was also a good correlation with the Kamlet–Taft β parameter, though it was best with the donor number.

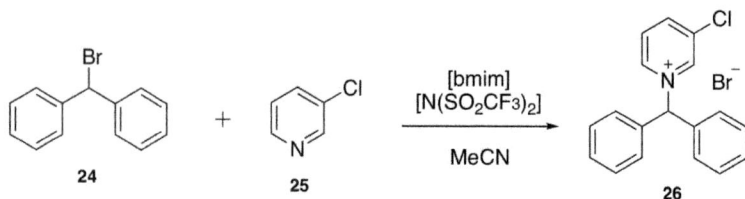

Scheme 9 A unimolecular substitution process that has been examined in mixtures containing the ionic liquid [bmim][N(SO$_2$CF$_3$)$_2$].[45]

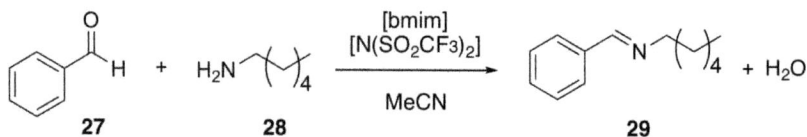

Scheme 10 A bimolecular condensation process that has been examined in mixtures containing the ionic liquid [bmim][N(SO$_2$CF$_3$)$_2$].[51,52]

for related processes with significant charge development in the transition state that the ionic liquid will order about and stabilize that transition state; at low proportions of salt in the reaction mixture, a rate-constant enhancement might be expected, while entropic penalties for solvent reorganization would result in that rate constant decreasing at greater proportions of the ionic liquid in the reaction mixture. Such qualitative prediction has been demonstrated to be effective for the unimolecular component of the substitution reaction of the bromide **24** by the pyridine **25** to give the salt **26** (Scheme 9).[45,47,75,r]

Such predictability would be notably more useful if it were quantitative; for example, some indication of the magnitude of the rate-constant change and the solvent composition at which this would occur would be beneficial. Unsurprisingly, the closer the reaction parallels one that has already been studied, the better the predictability of the model.

For example, consider the reaction between the aldehyde **27** and the amine **28** (Scheme 10)[51–53]; it involves a nitrogen nucleophile and some degree of charge development in the transition state. As such, it parallels previous studies on bimolecular nucleophilic substitution processes[46,47,75,76,] and it might be expected that there would be a rate-constant enhancement in the presence of an ionic liquid, with a mole-fraction dependence

---

[r] Unsurprisingly, the bimolecular component saw a rate constant enhancement consistent with that described previously for S$_N$2 reactions of pyridine **11** and benzyl bromide **10**.[47,75]

similar to that seen previously (Fig. 3 above). The origin of the rate-constant enhancement would be expected to be a key interaction between the charged center on the nitrogen nucleophile **28** and the charged center on the cation of the ionic liquid, which is disrupted on moving to the transition state resulting in an entropic benefit.

Generally, these predictions hold up remarkably well. The rate-constant profile with changing solvent composition for this reaction is shown in Fig. 6, and it is very similar to that shown previously (Fig. 3); there is an initial rapid increase in rate constant, and the slope of the plot decreases with increasing proportion of salt in the reaction mixture. The activation parameters show an enthalpic cost overcome by an entropic benefit,[51] consistent with organization about, and stabilization of the starting materials. Changing each of the components of the ionic liquid[51,52] gives changes in activation parameters consistent with key interaction being between the cation of the ionic liquid and the nucleophilic nitrogen center on the amine **28**.

While this predictability is effective, there are subtleties that are not readily envisaged. Predicting the magnitude of the rate-constant enhancement is not straightforward. While the dependence of the rate constant on solvent composition is similar in each case, there are some differences such as the

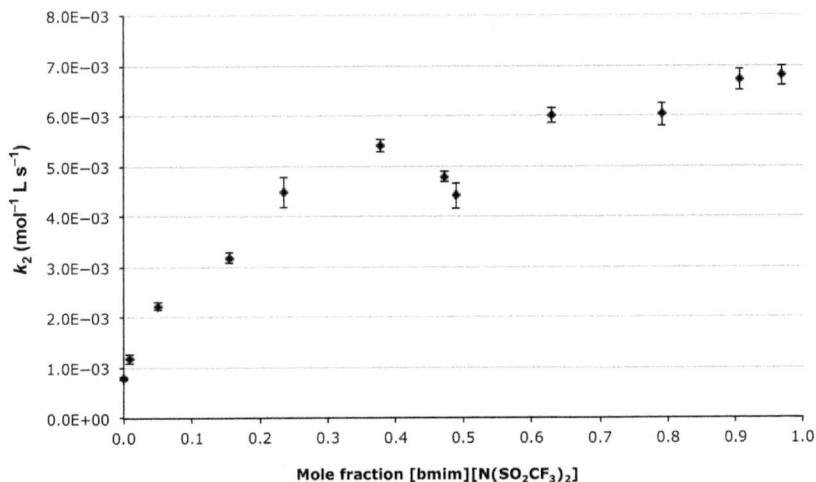

Fig. 6 The bimolecular rate constant for the reaction of the amine **28** with benzaldehyde **27** in different mole fractions of [bmim][N(SO$_2$CF$_3$)$_2$] in acetonitrile. *Reproduced from Keaveney ST, Schaffarczyk McHale KS, Haines RS, Harper JB. Developing principles for predicting ionic liquid effects on reaction outcome. A demonstration using a simple condensation reaction. Org Biomol Chem. 2014;12:7092–7099 with permission from the Royal Society of Chemistry.*

"dip" that is clearly visible at χ_{salt} ca. 0.45.[s] This deviation has been seen in other measurements and might be attributed to a change in solvent composition from "ionic liquid dissolved in molecular solvent" to "molecular solvent dissolved in ionic liquid"[99,100]; the extent to which this affects a given reaction is not readily predicted. Finally, while the changes in activation parameters on changing the components of the ionic liquid are as might be expected, the subtle balance in enthalpic and entropic effects means that the order of rate constants in different solvent mixtures does not parallel that for the bimolecular substitution case[77]; this lack of correlation is despite the fact that there are good correlations between the observed activation parameters and the Kamlet–Taft β parameter of the ionic liquid.[52]

Perhaps unsurprisingly, the most effective predictability comes when the reaction closely mirrors one that has been examined in ionic liquids. An example is the ethanolysis of the fluoropyridine **30** (Scheme 11)[101]; given the nature of the reaction, any ionic liquid solvent effects would be expected to be similar to the corresponding ethanolysis of the fluorobenzene **13** (Scheme 5). A rate-constant enhancement would be expected compared to the molecular solvent, ethanol, due to a combination of interactions at the microscopic level between the ionic liquid and the starting material **30**, and the ionic liquid and the transition state.

For this example, the predictive framework was remarkably effective! Not only is there a rate-constant enhancement with the same trend as would be expected from the similar nucleophilic aromatic substitution reaction, but the results show the magnitude of the rate-constant enhancement correlates well with the fluorobenzene **13** case.[101] This feature is best observed in

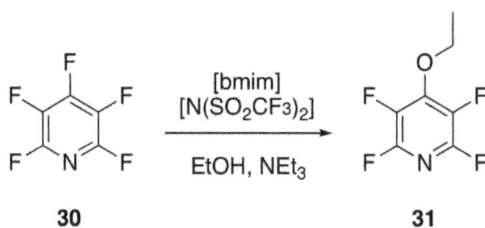

30   **31**

Scheme 11 A nucleophilic aromatic substitution process that has been examined in mixtures containing one of a range of ionic liquids.[101]

---

[s] The dip may be present in the substitution case, but the reduced change in the rate constant may result in it not being clearly observed.

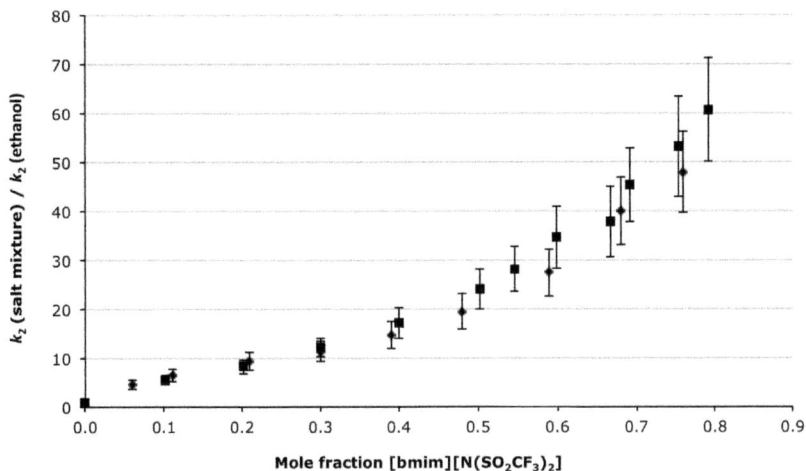

Fig. 7 A plot of the rate constants for the ethanolysis of each of the electrophiles pentafluoropyridine **30** (◆) and fluorodinitrobenzene **13** (■) in mixtures containing different proportions of [bmim][N(SO$_2$CF$_3$)$_2$] in ethanol, normalized to the rate constant for the same reaction in ethanol. *Adapted from Hawker RR, Haines RS, Harper JB. Predicting solvent effects in ionic liquids: extension of a nucleophilic aromatic substitution reaction on a benzene to a pyridine.* J Phys Org Chem. *2018. https://doi.org/10.1002/poc.3862 with permission from John Wiley & Sons.*

Fig. 7, where the rate-constant enhancement relative to ethanol at varying proportions of the salt [bmim][N(SO$_2$CF$_3$)$_2$] in the reaction mixture for each of the reactions shown in Schemes 5 and 11 is given.[t] The microscopic origins of the rate-constant enhancement were also shown to be the same in each case—at low mole fractions of ionic liquid the rate constant is entropically driven and at higher mole fractions there is more of an enthalpic component—with the rate-constant enhancement being due to a combination of these effects with the ionic liquid—starting material interaction being dominant.[101]

3.2 Rational Design of an Ionic Liquid Solvent

Initially it is useful to define "design" of an ionic liquid in the context it will be used here. Here, design refers to choosing the components of the ionic liquid (based on some fundamental property of those component ions) in order to control reaction outcome through solvent effects. Importantly, akin to the discussion in the rest of this review, it does not rely on the

---

[t] Similar correlations were observed with the other ionic liquids considered.

incorporation of functionalities into the ionic liquid that take part in the reaction. This particular definition of solvent design in the context of ionic liquids was introduced by Welton[102] in studies into nucleophilic aromatic substitutions using aniline nucleophiles. In that example, while assessment of ionic liquid effects was based on isolated yield of products, solvents were designed based on the empirical observation that ionic liquids with basic anions promoted the reaction.

The above discussions have introduced the fact that ionic liquid solvent effects might be reasonably predicted through an understanding of the microscopic interactions responsible. Of interest would be to maximize these solvent effects through appropriate choice of the ionic components of the solvent. To some extent this potential has already been introduced above, with different ionic liquids having different effects on reaction outcome that can be explained in terms of the microscopic interactions; the "best" ionic liquid could then be used. However, of particular interest here is to rationally design an ionic liquid solvent, possibly even one that had not been prepared before, to give the reaction outcome you want.

Such rational design has already been demonstrated for the bimolecular nucleophilic substitution reaction presented in Scheme 4, where ionic liquids resulted in rate-constant enhancement across a wide range of reaction mixtures. This enhancement is due to an interaction between the charged center of the ionic liquid cation, with the lone pair on the nucleophile, pyridine **11**.[75,76] It was envisaged that this interaction could be increased (and hence the solvent effect increased) through increasing accessibility to the charged center and by making the charged center more electron deficient. At the same time, the use of a particularly non-coordinating anion (which would not compete with pyridine **11** for interactions with the cation) would be necessary; for this reason, throughout these studies $[N(SO_2CF_3)_2]^-$ was used.

This concept was initially investigated by designing ionic liquids with electronegative chlorine atoms added to the imidazolium ring of the ionic liquid cation; the effect of the ionic liquids containing these cations was compared with the ionic liquids containing cations with methyl groups at the equivalent positions.[103] These substituents are of comparable size, but differing electronic character, and the rate-constant enhancement for the reaction of pyridine **11** with benzyl bromide **10** seen on moving from $[bm_3im][N(SO_2CF_3)_2]$ to $[b45Cl_2mim][N(SO_2CF_3)_2]$ (cations shown in Fig. 8, relative rate-constant data shown in Table 3) is consistent with

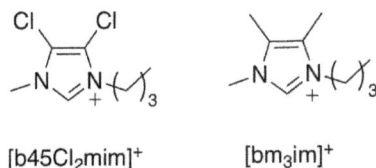

[b45Cl₂mim]⁺ [bm₃im]⁺

Fig. 8 An example of a chloroimidazolium-based cation that was incorporated in an ionic liquid designed to enhance the bimolecular nucleophilic substitution reaction shown in Scheme 4, along with the methyl analogue.[103] The abbreviation for the methyl variant shown is the one that is used in literature; it is noted that the abbreviation itself does not specify the site of methylation.

Table 3 The Rate Constant for the Reaction Shown in Scheme 4 at 292.2 K in Each of a Series of Ionic Liquids at Mole Fraction Given, Relative to Acetonitrile

Solvent	x_{salt}	k_{salt}/k_{CH3CN}
$[bm_3im][N(SO_2CF_3)_2]^a$	0.85	1.1 ± 0.1
$[b45Cl_2mim][N(SO_2CF_3)_2]^b$	0.85	2.80 ± 0.06
$[hpy][N(SO_2CF_3)_2]^c$	0.36	2.19 ± 0.06
$[bmim][N(SO_2CF_3)_2]^d$	0.86	2.39 ± 0.09
$[(mim)_2p_e]_2[N(SO_2CF_3)_2]^c$	0.78	3.6 ± 0.2
$[Fhpy][N(SO_2CF_3)_2]^c$	0.36	4.3 ± 0.1
$[bmmo][N(SO_2CF_3)_2]^c$	0.70	5.4 ± 0.5
$[btl][N(SO_2CF_3)_2]^c$	0.85	7 ± 1

[a]Data from Tanner et al.[77]
[b]Data from Hawker et al.[103]
[c]Data from Hawker et al.[48]
[d]Data from Yau et al.[75]

greater interaction with as a result of the electron–withdrawing nature of the chloro substituents. It should be noted that two such substituents were required for a difference to be seen in the rate constants.

After this initial proof of principle, a greater range of ionic liquids were envisaged using the same criteria as above; representative examples of the cations considered are shown in Fig. 9. Each of these ionic liquid cations had an enhanced interaction with the pyridine **11** leading to entropically driven rate-constant enhancements in ionic liquids made up of these cations and the $[N(SO_2CF_3)_2]^-$ anion (Table 3).[48] From the results presented, the "best" ionic liquid for this reaction was determined through consideration

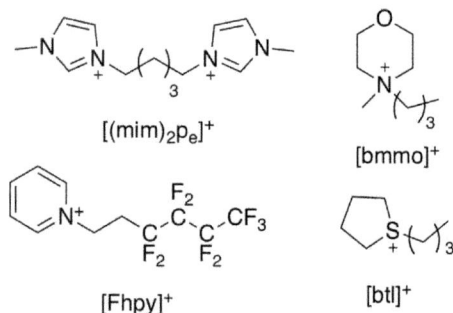

[(mim)$_2$p$_e$]$^+$

[bmmo]$^+$

[Fhpy]$^+$

[btl]$^+$

Fig. 9 Examples of cations that was incorporated into ionic liquid designed to enhance the bimolecular nucleophilic substitution reaction shown in Scheme 4.[48]

of the rate-constant enhancement, cost and ease of synthesis of the ionic liquid. Both of the ionic liquids [bmmo][N(SO$_2$CF$_3$)$_2$] and [(mim)$_2$p$_e$][N (SO$_2$CF$_3$)$_2$]$_2$ were chosen based on the above categories.[48]

While only demonstrated for one reaction, the ability to design a solvent to maximize the interactions responsible for ionic liquid solvent effects should be transferable, not just to related systems (such as the condensation reaction in Scheme 10, though it would be expected that the ionic liquids demonstrated to be effective above would also be effective in that case) but any examples of reactions where the microscopic interactions are understood.

While the examples above demonstrate that there is the potential to design ionic liquids to control reaction outcome, the predictive method is at best qualitative. This level of predictability is not necessarily poor, given that the equivalent predictive framework for molecular solvents is similarly qualitative.[69] However, it would be advantageous to be able to quantify the effects.

Note that in the design of the ionic liquids to control reaction outcome discussed here the key point is to incorporate the interactions demonstrated to be responsible for the rate-constant enhancements. This introduces a manner in which it might be possible to quantify the changes in reaction rate constants; if the rate constant could be correlated to the strength of a key interaction, measuring such an interaction would allow the ionic liquid effect to be predicted. At this stage, such quantitation has not been undertaken and indirect measures of solvent parameters (introduced above) have limited efficacy. Computational techniques seem the logical starting point for such and are currently underway; importantly such studies would mean that potential ionic liquid solvents could be screened prior to synthesis.

In addition, experimental techniques, such as determining the diffusion of a solute[99] (which might be considered an indirect measurement but is at least directly related to the solute molecule, rather than a somewhat arbitrary probe molecule that is necessary for determining solvent parameters) and measuring activity coefficients[u] (again an indirect measure of solute–solvent interaction), may provide assessment of potential interactions once the ionic liquids had been prepared.[104]

Before concluding this section, it is worthwhile briefly considering that the solvent that maximizes the rate constant (or other reaction outcome) might be problematic to use for another reason (or reasons). These reasons include an inability to dissolve the reagents and being solid at the temperature being used (see, for example, rational choice of solvent for the aromatic substitution process shown in Scheme 5[50]). A potential way to overcome this would be to consider using an ionic liquid containing either more than one type of cation, more than one type of anion, or both; such examples might be considered a mixture of salts. Use of such mixtures with different anions as solvents for the ethanolysis of the benzene **13** has already been demonstrated.[105] It should be noted that the preferential formation of ion pairs between given components of the mixture needs to be considered; mixtures containing very similar cations (such as those proposed by Davis[106]) might alleviate this problem in cases where lowering the melting point is desired.

3.3 Biasing Product Selectivity

The above arguments in this section have focused on exploiting ionic liquid solvents to increase rate constants; reaction mechanisms have been considered in isolation. When considering them together there brings the possibility that an ionic liquid might be used to favor one process relative to another, resulting in a different selectivity for the reaction. (It is important to note that this selectivity is subtly different to the selectivities already observed in ionic liquids.) Of particular interest is if the two processes have markedly different dependencies on the proportion of the ionic liquid in the reaction mixture; it might be possible to use the same ionic liquid (albeit different amounts of the same ionic liquid) to favor one process under one set of conditions and the other process under another set of conditions.

---

[u] Some correlation involving these measures and rate constants for pericyclic processes has already been demonstrated.[104]

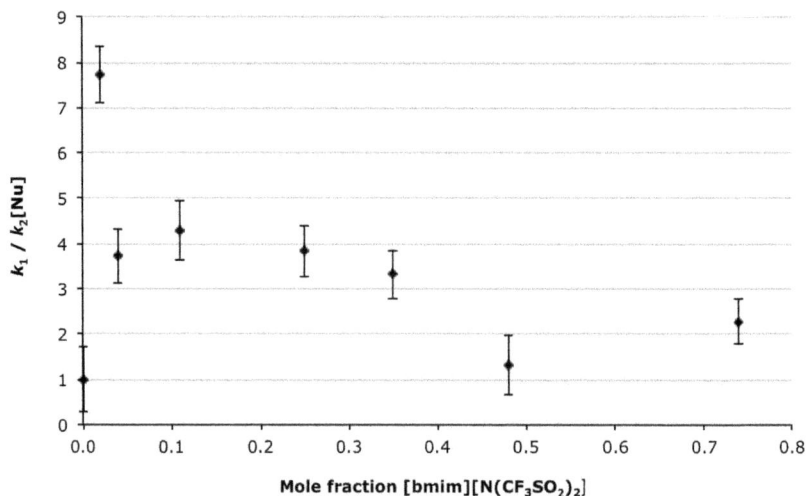

Fig. 10 The ratio between k_1 and $k_2[Nu]$ for the reaction between bromodiphenylmethane **24** and 3-chloropyridine **25** across various mole fractions of [bmim][N(SO$_2$CF$_3$)$_2$] in acetonitrile. *Reproduced from Keaveney ST, White BP, Haines RS, Harper JB. The effects of an ionic liquid on unimolecular substitution processes: the importance of the extent of transition state solvation. Org Biomol Chem. 2016;14:2572–2580 with permission from the Royal Society of Chemistry.*

An introductory example of this principle is the situation where one product can be formed through two different reaction mechanisms from the same reagents. In the extreme, one mechanism occurs in molecular solvents and another when any amount of an ionic liquid is present. There is also the potential for a more gradual change, with the proportion of a reaction proceeding through each mechanism varying as the amount of an ionic liquid in the reaction mixture changes. The former is exemplified in reactions that proceed through either E2 (in ionic liquids) or E1cB (in molecular solvents) mechanisms[107,108] while the latter is demonstrated by the already introduced conversion of the benzyl bromide **24** to the salt **26** (Scheme 9; Fig. 10).[45]

The latter is a particularly illustrative example as the unimolecular and bimolecular substitution pathways have notably different dependencies on the amount of ionic liquid [bmim][N(SO$_2$CF$_3$)$_2$] present in the reaction mixture. As such, the proportion of product that forms through each process varies with the amount of ionic liquid; this variation is illustrated in Fig. 10. In and of itself, whether the product **26** is formed through either a unimolecular or a bimolecular process is irrelevant. However, this does demonstrate that if two separate reactions proceeding through each of these

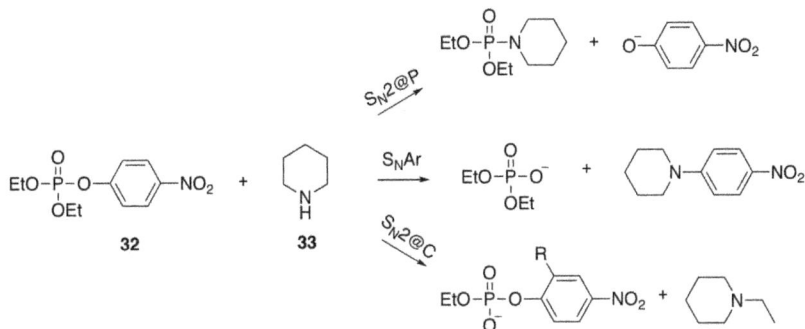

Scheme 12 The reaction between the phosphate triester **32** and piperidine **33**, which has been studied in a series of ionic liquids. Reaction can occur through three different pathways: S_N2 at the phosphorus center, S_NAr at the aromatic moiety, and S_N2 at the aliphatic carbon.[109–111]

mechanisms (these may be in the same molecule or in different molecules) were occurring, then the ratio of products might be controlled.

Such control of the selectivity of reactions has been demonstrated extensively by Pavez et al.[109–111] in the reaction of phosphorus derivatives such as Paraoxon **32**; the possible reactions are shown with piperidine **33** in Scheme 12. These three pathways are (1) S_N2 at the phosphorus center, (2) S_NAr at the aromatic carbon, and (3) S_N2 at the aliphatic carbon. While carried out in a single solvent composition (generally χ_{salt} ca. 0.45), these types of reactions have been studied in a number of imidazolium-, ammonium-, and pyrrolidinium-based ionic liquids. The product distribution did vary with the phosphorus substrate and the nature of the ionic liquid, generally the bimolecular process at phosphorus was favored in the ionic liquid. While no dependence of reaction outcome on proportion of ionic liquid in the reaction mixture has been determined, the known dependence of the rate constants for a related $S_N2@P$ processes[54–56] suggests that the mole fraction chosen might be ideal for maximizing the selectivity of this process. The rate of reaction for this process in the case of the substrate **32** correlates well with the Kamlet–Taft β parameter of the solvent, again consistent with other reports,[56] though lack of the same correlation in the case of the other pathways means the selectivity cannot be correlated with this solvent parameter.

Another example of selectivity in product formation, this time examined over different proportions of ionic liquid in the reaction mixtures, was competing substitution and elimination reactions on the phenethyl chloride **34** (Scheme 13).[46] This process is complex, given that each of the products **35**

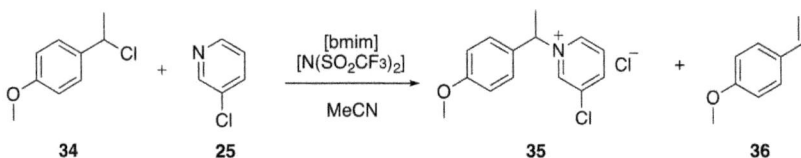

Scheme 13 The reaction between the chloride **34** and pyridine **25** which has been studied in mixtures containing the ionic liquid [bmim][N(SO$_2$CF$_3$)$_2$].[46]

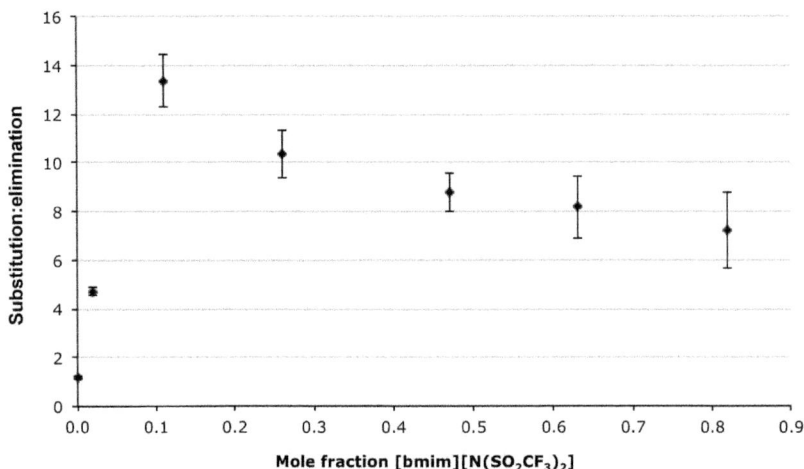

Fig. 11 The ratio between the substitution product **35** and the elimination product **36** for the reaction between the chloride **34** and 3-chloropyridine **25** (ca. 0.2 mol L^{-1}) across various mole fractions of [bmim][N(SO$_2$CF$_3$)$_2$] in acetonitrile at 314 K. *Reproduced from Keaveney ST, Harper JB. Towards reaction control using an ionic liquid: biasing outcomes of reactions of benzyl halides. RSC Adv. 2013;3:15698–15704 with permission from the Centre National de la Recherche Scientifique (CNRS) and the Royal Society of Chemistry.*

and **36** is formed through unimolecular and bimolecular processes, but kinetic analyses of similar mechanisms introduced earlier can be used to interpret the preference for substitution in mixtures containing [bmim][N(SO$_2$CF$_3$)$_2$] shown in Fig. 11.

While these simple cases demonstrate the potential for selectivity based on the different effects of ionic liquids on different reaction mechanisms, there is clearly more scope for more synthetic applications. It should also be noted that these selectivities might be further enhanced through varying the temperature, particularly when processes are affected significantly differently by ionic liquids in terms of entropy of activation (for example, compare S$_N$1 and S$_N$2 processes).

4. SUMMARY AND OUTLOOK

Twenty years ago ionic liquids were something of a poorly understood curiosity and, while the potential to use them as solvents had been identified, there was a distinct lack of understanding as to their solvent effects on reaction outcome. As outlined herein, since that time there has been a great deal of progress made and initial, sometimes erroneous, assumptions have been challenged and corrected. Importantly, a variety of ways to consider solvent effects (both qualitative and quantitative) have been introduced that allow not only rationalization of solvent effects observed but also the design of solvents to control reaction outcome.

It is interesting to consider the ongoing direction of this research area. While the understanding developed is significant, it certainly is not comprehensive and further investigations are needed across a greater range of reaction types. As this understanding has broadened, so has the range of ionic liquids available and their usage in combination with other solvents (for a recent commentary on the use of mixtures of ionic liquids, and other salts, with molecular solvents in areas further afield than those described here, see MacFarlane et al.[112]). Application of this system remains the biggest challenge and with solvent effects more readily predicted, concern has turned to cost; can these salts be made on sufficient scale and recycled effectively so that the process is economically viable? (For interesting work on this problem in the context of a biorefinery, see Hallett et al.[113]) It might be contended that, rather than considering processes where molecular solvents could be replaced with ionic liquids, processes be targeted that are not viable in molecular solvents but may be so in mixtures containing an ionic liquid; that is, to develop new processes rather than modify old ones.

ACKNOWLEDGMENTS

J.B.H. would like to acknowledge, particularly, all of the coworkers who have worked with him on ionic liquid projects over the last 15 years. Without their hard work, much of work described in this review would not have been carried out. Financial support from the Australian Research Council Discovery Project Funding Scheme (Projects DP130102331 and DP180103682) and the University of New South Wales (Goldstar and Faculty Research Grant Programmes) is also gratefully acknowledged.

REFERENCES

1. El Abedin SZ, Endres F. Ionic liquids: the link to high-temperature molten salts? *Acc Chem Res.* 2007;40:1106–1113.
2. Welton T. Room-temperature ionic liquids. Solvents for synthesis and catalysis. *Chem Rev.* 1999;99:2071–2083.

3. Hussey CL. Room temperature haloaluminate ionic liquids. Novel solvents for transition metal solution chemistry. *Pure Appl Chem*. 1988;60:1763–1772.
4. Dubois RH, Zaworotko MJ, White PS. Complex hydrogen-bonded cations. X-ray crystal structure of [((C$_6$H$_5$CH$_2$)NH$_3$)$_4$Cl][AlCl$_4$]$_3$ and its relevance to the structure of basic chloroaluminate room-temperature melts. *Inorg Chem*. 1989;28:2019–2020.
5. Wilkes JS, Zaworotko MJ. Air and water stable 1-ethyl-3-methylimidazolium based ionic liquids. *J Chem Soc Chem Commun*. 1992;965–967.
6. Bonhôte P, Dias A-P, Papageorgiou N, Kalyanasundaram K, Grätzel M. Hydrophobic, highly conductive ambient-temperature molten salts. *Inorg Chem*. 1996;35:1168–1178.
7. Huddleston JG, Willauer HD, Swatloski RP, Visser AE, Rogers RD. Room temperature ionic liquids as novel media for 'clean' liquid-liquid extraction. *Chem Commun*. 1998;1765–1766.
8. Ramsay W. On picoline and its derivatives. *Philos Mag Ser*. 1876;5(2):269–281.
9. Walden P. Molecular weights and electrical conductivity of several fused salts. *Bull Russ Acad Sci*. 1914;8:405–422.
10. Wilkes JS. A short history of ionic liquids—from molten salts to neoteric solvents. *Green Chem*. 2002;4:73–80.
11. Holbrey JD, Seddon KR. Ionic liquids. *Clean Prod Process*. 1999;1:223–226.
12. Yau HM, Keaveney ST, Butler BJ, et al. Towards solvent-controlled reactivity in ionic liquids. *Pure Appl Chem*. 2013;85:1979–1990.
13. Hallett JP, Welton T. Room-temperature ionic liquids: solvents for synthesis and catalysis. 2. *Chem Rev*. 2011;111:3508–3576.
14. Seddon KR. Room-temperature ionic liquids: neoteric solvents for clean catalysis? *Kinet Catal*. 1996;37:693–697.
15. Swiderski K, McLean A, Gordon CM. Estimates of internal energies of vaporisation of some room temperature ionic liquids. *Chem Commun*. 2004;2178–2179.
16. Verevkin SP, Zaitsau DH, Emel'yanenko VN, et al. Making sense of enthalpy of vaporization trends for ionic liquids: new experimental and simulation data show a simple linear relationship and help reconcile previous data. *J Phys Chem B*. 2013;117:6473–6486.
17. Earle MJ, Esperança JMSS, Gilea MA, et al. The distillation and volatility of ionic liquids. *Nature*. 2006;439:831–834.
18. Jordan A, Gathergood N. Biodegradation of ionic liquids—a critical review. *Chem Soc Rev*. 2015;44:8200–8237.
19. Mai NL, Ahn K, Koo Y-M. Methods for recovery of ionic liquids. *Process Biochem*. 2014;49:872–881.
20. Wu B, Liu W, Zhang Y, Wang H. Do we understand the recyclability of ionic liquids? *Chem A Eur J*. 2009;15:1804–1810.
21. Fang S, Zhang Z, Jin Y, et al. New functionalized ionic liquids based on pyrrolidinium and piperidinium cations with two ether groups as electrolytes for lithium battery. *J Power Sources*. 2011;196:5637–5644.
22. Steinrück H-P, Wasserscheid P. Ionic liquids in catalysis. *Catal Lett*. 2015;145:380–397.
23. Zhang Q, Zhang S, Deng Y. Recent advances in ionic liquid catalysis. *Green Chem*. 2011;13:2619–2637.
24. Harper JB, Kobrak MN. Understanding organic processes in ionic liquids: achievements so far and challenges remaining. *Mini-Rev Org Chem*. 2006;3:253–259.
25. Earle MJ, Katdare SP, Seddon KR. Paradigm confirmed: the first use of ionic liquids to dramatically influence the outcome of chemical reactions. *Org Lett*. 2004;6:707–710.
26. Weber CC, Masters AF, Maschmeyer T. Structural features of ionic liquids: consequences for material preparation and organic reactivity. *Green Chem*. 2013;15:2655–2679.

27. Chatel G, MacFarlane DR. Ionic liquids and ultrasound in combination: synergies and challenges. *Chem Soc Rev.* 2014;43:8132–8149.
28. Hawker RR, Haines RS, Harper JB. Variation of the cation of ionic liquids: the effects on their physicochemical properties and reaction outcome. Società Chimica Italiano; 2014:141–213. Targets in Heterocyclic Systems; vol 18.
29. Keaveney ST, Haines RS, Harper JB. Reactions in ionic liquids. New Jersey: Wiley; 2017, Encyclopedia of Physical Organic Chemistry; vol 2.
30. Keaveney ST, Haines RS, Harper JB. Ionic liquid solvents: the importance of microscopic interactions in predicting organic reaction outcomes. *Pure Appl Chem.* 2017;89:745–757.
31. Gilbert A, Haines RS, Harper JB. Selecting ionic liquids to enhance and control reaction outcomes. In: Reedijk J, ed. *Elsevier Reference Module in Chemistry, Molecular Sciences and Chemical Engineering.* Waltham, MA: Elsevier; 2018: 26-Jul-18. https://doi.org/10.1016/B978-0-12-409547-2.14212-X.
32. Giernoth R. Task-specific ionic liquids. *Angew Chem Int Ed.* 2010;49:2834–2839.
33. Alonso DA, Baeza A, Chinchilla R, Guillena G, Pastor IM, Ramón DJ. Deep eutectic solvents: the organic reaction medium of the century. *Eur J Org Chem.* 2016;612–632.
34. Ueno K, Yoshida K, Tsuchiya M, Tachikawa N, Dokko K, Watanabe M. Glyme-lithium salt equimolar molten mixtures: concentrated solutions or solvate ionic liquids? *J Phys Chem B.* 2012;116:11323–11331.
35. Obregón-Zúñiga A, Milán M, Juaristi E. Improving the catalytic performance of (S)-proline as organocatalyst in asymmetric aldol reactions in the presence of solvate ionic liquids: involvement of a supramolecular aggregate. *Org Lett.* 2017;19:1108–1111.
36. Eyckens DJ, Champion ME, Fox BL, et al. Solvate ionic liquids as reaction media for electrocyclic transformations. *Eur J Org Chem.* 2016;913–917.
37. Eyckens DJ, Henderson LC. Synthesis of α-aminophosphonates using solvate ionic liquids. *RSC Adv.* 2017;7:27900–27904.
38. Fischer T, Sethi A, Welton T, Woolf J. Diels-Alder reactions in room-temperature ionic liquids. *Tetrahedron Lett.* 1999;40:793–796.
39. Dzyuba SV, Bartsch RA. Expanding the polarity range of ionic liquids. *Tetrahedron Lett.* 2002;43:4657–4659.
40. Aggarwal A, Lancaster NL, Sethi AR, Welton T. The role of hydrogen bonding in controlling the selectivity of Diels-Alder reactions in room-temperature ionic liquids. *Green Chem.* 2002;4:517–520.
41. Vidiš A, Ohlin CA, Laurenczy G, Küsters E, Sedelmeier G, Dyson PJ. Rationalisation of solvent effects in the Diels-Alder reaction between cyclopentadiene and methyl acrylate in room temperature ionic liquids. *Adv Synth Catal.* 2005;347:266–274.
42. Man BYW, Hook JM, Harper JB. Substitution reactions in ionic liquids. A kinetic study. *Tetrahedron Lett.* 2005;46:7641–7645.
43. Creary X, Willis ED, Gagnon M. Carbocation-forming reactions in ionic liquids. *J Am Chem Soc.* 2005;127:18114–18120.
44. D'Anna F, Frenna V, La Marca S, Noto R, Pace V, Spinelli D. On the characterization of some [bmim][X]/co-solvent binary mixtures: a multidisciplinary approach by using kinetic, spectrophotometric and conductometric investigations. *Tetrahedron.* 2008;64:672–680.
45. Keaveney ST, White BP, Haines RS, Harper JB. The effects of an ionic liquid on unimolecular substitution processes: the importance of the extent of transition state solvation. *Org Biomol Chem.* 2016;14:2572–2580.
46. Keaveney ST, Harper JB. Towards reaction control using an ionic liquid: biasing outcomes of reactions of benzyl halides. *RSC Adv.* 2013;3:15698–15704.

47. Schaffarczyk McHale KS, Hawker RR, Harper JB. Nitrogen versus phosphorus nucleophiles—how changing the nucleophilic heteroatom affects ionic liquid solvent effects in bimolecular nucleophilic substitution processes. *New J Chem.* 2016;40: 7437–7444.
48. Hawker RR, Haines RS, Harper JB. Rational selection of the cation of an ionic liquid to control the reaction outcome of a substitution reaction. *Chem Commun.* 2018;54:2296–2299.
49. Hawker RR, Wong MJ, Haines RS, Harper JB. Rationalising the effects of ionic liquids on a nucleophilic aromatic substitution reaction. *Org Biomol Chem.* 2017;15:6433–6440.
50. Hawker RR, Haines RS, Harper JB. The effect of varying the anion of an ionic liquid on the solvent effects on a nucleophilic aromatic substitution reaction. *Org Biomol Chem.* 2018;16:3453–3464.
51. Keaveney ST, Schaffarczyk McHale KS, Haines RS, Harper JB. Developing principles for predicting ionic liquid effects on reaction outcome. A demonstration using a simple condensation reaction. *Org Biomol Chem.* 2014;12:7092–7099.
52. Keaveney ST, Haines RS, Harper JB. Developing principles for predicting ionic liquid effects on reaction outcome. The importance of the anion in controlling microscopic interactions. *Org Biomol Chem.* 2015;13:3771–3780.
53. Keaveney ST, Haines RS, Harper JB. Ionic liquid effects on a multistep process. Increased product formation due to enhancement of all steps. *Org Biomol Chem.* 2015;13:8925–8936.
54. Butler BJ, Harper JB. The effect of an ionic liquid on the rate of reaction at a phosphorus centre. *New J Chem.* 2015;39:213–219.
55. Butler BJ, Harper JB. The effect of the structure of the cation of an ionic liquid on the rate of reaction at a phosphorus centre. *J Phys Org Chem.* 2016;29:700–708.
56. Butler BJ, Harper JB. The effect of the structure of the anion of an ionic liquid on the rate of reaction at a phosphorus centre. *J Phys Org Chem.* 2018. https://doi.org/10.1002/poc.3819.
57. Khupse ND, Kumar A. The cosolvent-directed Diels-Alder reaction in ionic liquids. *J Phys Chem A.* 2011;115:10211–10217.
58. Cammarata L, Kazarian SG, Salter PA, Welton T. Molecular states of water in room temperature ionic liquids. *Phys Chem Chem Phys.* 2001;3:5192–5200.
59. Seddon KR, Stark A, Torres MJ. Influence of chloride, water and organic solvent on the physical properties of ionic liquids. *Pure Appl Chem.* 2000;72:2275–2287.
60. George SRD, Edwards GL, Harper JB. The effects of ionic liquids on azide-alkyne cycloaddition reactions. *Org Biomol Chem.* 2010;8:5354–5358.
61. Pan Y, Boyd LE, Kruplak JF, Cleland WE, Wilkes JS, Hussey CL. Physical and transport properties of bis(trifluoromethylsulfonyl)imide-based room-temperature ionic liquids: application to the diffusion of tris(2,2′-bipyridyl)ruthenium(II). *J Electrochem Soc.* 2011;158:f1–f9.
62. Pindur U, Lutz G, Acceleration OC. Selectivity enhancement of Diels-Alder reactions by special and catalytic methods. *Chem Rev.* 1992;93:741–761.
63. Lee CW. Diels-Alder reactions in chloroaluminate ionic liquids: acceleration and selectivity enhancement. *Tetrahedron Lett.* 1999;40:2461–2463.
64. Kumar A, Pawar SS. Converting *exo*-selective Diels—Alder reaction to *endo*-selective in chloroloaluminate ionic liquids. *J Org Chem.* 2004;69:1419–1420.
65. Kumar A, Pawar SS. Ionic liquids as powerful solvent media for improving catalytic performance of silyl borate catalyst to promote Diels—Alder reactions. *J Org Chem.* 2007;72:8111–8114.
66. Ludley P, Karodia N. Phosphonium tosylates as solvents for the Diels-Alder reaction with 1,3-cyclopentadiene. *ARKIVOC.* 2002;172–175.

67. Rosella CE, Harper JB. The effect of ionic liquids on the outcome of nitrile oxide cycloadditions. *Tetrahedron Lett.* 2009;50:992–994.
68. Sheldon R. Catalytic reactions in ionic liquids. *Chem Commun.* 2001;2399–2407.
69. Smith MB, March J. *March's Advanced Organic Chemistry.* 5th ed. New York: Wiley-Interscience; 2001.
70. Crowhurst L, Lancaster NL, Arlandis JMP, Welton T. Manipulating solute nucleophilicity with room temperature ionic liquids. *J Am Chem Soc.* 2004;126:11549–11555.
71. Lancaster NL, Welton T, Young GB. A study of halide nucleophilicity in ionic liquids. *J Chem Soc Perkin Trans 2.* 2001;2267–2270.
72. Lancaster NL, Salter PA, Welton T, Young GB. Nucleophilicity in ionic liquids. 2.1 cation effects on halide nucleophilicity in a series of Bis(trifluoromethylsulfonyl)imide ionic liquids. *J Org Chem.* 2002;67:8855–8861.
73. Lancaster NL, Welton T. Nucleophilicity in ionic liquids. 3.1 anion effects on halide nucleophilicity in a series of 1-Butyl-3-methylimidazolium ionic liquids. *J Org Chem.* 2004;69:5986–5992.
74. Crowhurst L, Falcone R, Lancaster NL, Llopis-Mestre V, Welton T. Using Kamlet–Taft solvent descriptors to explain the reactivity of anionic nucleophiles in ionic liquids. *J Org Chem.* 2006;71:8847–8853.
75. Yau HM, Howe AG, Hook JM, Croft AK, Harper JB. Solvent reorganisation as the driving force for rate changes of Menshutkin reactions in an ionic liquid. *Org Biomol Chem.* 2009;7:3572–3575.
76. Yau HM, Croft AK, Harper JB. Investigating the origin of entropy-derived rate accelerations in ionic liquids. *Faraday Discuss.* 2012;154:365–371.
77. Tanner EEL, Yau HM, Hawker RR, Croft AK, Harper JB. Does the cation really matter? The effect of modifying an ionic liquid cation on an S_N2 process. *Org Biomol Chem.* 2013;11:6170–6175.
78. Keaveney ST, Francis DV, Cao W, Haines RS, Harper JB. Effect of modifying the anion of an ionic liquid on the outcome of an S_N2 process. *Aust J Chem.* 2015;68:31–35.
79. Yau HM, Barnes SA, Hook JM, Youngs TGA, Croft AK, Harper JB. The importance of solvent reorganisation in the effect of an ionic liquid on a unimolecular substitution process. *Chem Commun.* 2008;3576–3578.
80. D'Anna F, Frenna V, Noto R, Pace V, Spinelli D. Study of aromatic nucleophilic substitution with amines on nitrothiophenes in room-temperature ionic liquids: are the different effects on the behavior of Para-like and ortho-like isomers on going from conventional solvents to room-temperature ionic liquids related to solvation effects? *J Org Chem.* 2006;71:5144–5150.
81. Jones SG, Yau HM, Davies E, et al. Ionic liquids through the looking glass: theory mirrors experiment and provides further insight into aromatic substitution processes. *Phys Chem Chem Phys.* 2010;12:1873–1878.
82. D'Anna F, Marca SL, Noto R. Kemp elimination: a probe reaction to study ionic liquids properties. *J Org Chem.* 2008;73:3397–3403.
83. Yau HM, Chan SJ, George SRD, Hook JM, Croft AK, Harper JB. Ionic liquids: just molten salts after all? *Molecules.* 2009;14:2521–2534.
84. A. Wilkinson, McNaught AD. IUPAC. Compendium of chemical terminology. 2nd ed. (t"Gold Book"). Oxford: Blackwell Scientific Publications; 1997.
85. Isaacs NS. *Physical Organic Chemistry.* 2nd ed. Harlow: Addison Wesley Longman Limited; 1998.
86. Saha S, H-o H. Effect of water on the molecular structure and arrangement of nitrile functionalized ionic liquids. *J Phys Chem B.* 2006;110:2777–2781.
87. Kobrak MN, Lin H. Electrostatic interactions in ionic liquids: the dangers of dipole and dielectric descriptions. *Phys Chem Chem Phys.* 2010;12:1922–1932.

88. Welton T. Polarity of ionic liquids. In: Wasserscheid P, Welton T, eds. *Ionic Liquids in Synthesis*. 2nd ed. Weinheim, Germany: Wiley-VCH; 2008:130–175.

89. Jessop PG, Jessop DA, Fu D, Phan L. Solvatochromic parameters for solvents of interest in green chemistry. *Green Chem*. 2012;14:1245–1259.

90. Taft RW, Kamlet MJ. The solvatochromic comparison method. 2. The α-scale of solvent hydrogen-bond donor (HBD) acidities. *J Am Chem Soc*. 1976;98:2886–2894.

91. Kamlet MJ, Taft RW. The solvatochromic comparison method. I. the β-scale of solvent hydrogen-bond acceptor (HBA) basicities. *J Am Chem Soc*. 1976;98:377–383.

92. Kamlet MJ, Abboud JL, Taft RW. The solvatochromic comparison method. 6. The π^* scale of solvent polarities. *J Am Chem Soc*. 1977;99:6027–6039.

93. Bini R, Chiappe C, Mestre VL, Pomelli CS, Welton T. A rationalization of the solvent effect on the Diels-Alder reaction in ionic liquids using multiparameter linear solvation energy relationships. *Org Biomol Chem*. 2008;6:2522–2529.

94. Tiwari S, Khupse N, Kumar A. Intramolecular Diels – Alder reaction in ionic liquids: effect of ion-specific solvent friction. *J Org Chem*. 2008;73:9075–9083.

95. Bini R, Chiappe C, Mestre V, Pomelli C, Welton T. A theoretical study of the solvent effect on Diels-Alder reaction in room temperature ionic liquids using a supermolecular approach. *Theor Chem Acc*. 2009;123:347–352.

96. Chiappe C, Malvaldi M, Pomelli CS. Ab initio study of the Diels – Alder reaction of cyclopentadiene with acrolein in a ionic liquid by KS-DFT/3D-RISM-KH theory. *J Chem Theory Comput*. 2009;6:179–183.

97. Gutmann V, Wychera E. Coordination reactions in nonaqueous solutions. The role of the donor strength. *Inorg Nucl Chem Lett*. 1966;2:257–260.

98. Alarcón-Espósito J, Contreras R, Tapia RA, Campodónico PR. Gutmann's donor numbers correctly assess the effect of the solvent on the kinetics of S_NAr reactions in ionic liquids. *Chem A Eur J*. 2016;22:13347–13351.

99. Keaveney ST, Schaffarczyk Mc Hale KS, Stranger JW, Ganbold B, Price WS, Harper JB. NMR diffusion measurements as a simple method to examine solvent–solvent and solvent–solute interactions in mixtures of the ionic liquid [Bmim][N(SO$_2$CF$_3$)$_2$] and acetonitrile. *Chemphyschem*. 2016;17:3853–3862.

100. Keaveney ST, Greaves TL, Kennedy DF, Harper JB. Understanding the effect of solvent structure on organic reaction outcomes when using ionic liquid/acetonitrile mixtures. *J Phys Chem B*. 2016;120:12687–12699.

101. Hawker RR, Haines RS, Harper JB. Predicting solvent effects in ionic liquids: extension of a nucleophilic aromatic substitution reaction on a benzene to a pyridine. *J Phys Org Chem*. 2018;31:e3862. https://doi.org/10.1002/poc.3862.

102. Newington I, Perez-Arlandis JM, Welton T. Ionic liquids as designer solvents for nucleophilic aromatic substitutions. *Org Lett*. 2007;9:5247–5250.

103. Hawker RR, Panchompoo J, Aldous L, Harper JB. Novel chloroimidazolium-based ionic liquids: synthesis, characterisation and behaviour as solvents to control reaction outcome. *ChemPlusChem*. 2016;81:574–583.

104. Keaveney ST, Haines RS, Harper JB. Investigating solvent effects of an ionic liquid on pericyclic reactions through kinetic analyses of simple rearrangements. *ChemPlusChem*. 2017;82:449–457.

105. Tanner EEL, Hawker RR, Yau HM, Croft AK, Harper JB. Probing the importance of ionic liquid structure: a general ionic liquid effect on an S_NAr process. *Org Biomol Chem*. 2013;11:7516–7521.

106. Stenson AC, West KN, Reichert WM, et al. Multi-ion ionic liquids and a direct, reproducible, diversity-oriented way to make them. *Chem Commun*. 2015;51:15914–15916.

107. D'Anna F, Frenna V, Pace V, Noto R. Effect of ionic liquid organizing ability and amine structure on the rate and mechanism of base induced elimination of 1,1,1-tribromo-2,2-bis(phenyl-substituted)ethanes. *Tetrahedron*. 2006;62:1690–1698.

108. Allen C, Sambasivarao SV, Acevedo O. An ionic liquid dependent mechanism for base catalyzed β-elimination reactions from QM/MM simulations. *J Am Chem Soc*. 2013;135:1065–1072.

109. Pavez P, Millan D, Cocq C, Santos JG, Nome F. Ionic liquids: anion effect on the reaction of *O,O*-diethyl *O*-(2,4-dinitrophenyl) phosphate triester with piperidine. *New J Chem*. 2015;39:1953–1959.

110. Pavez P, Millán D, Morales JI, Castro EA, López AC, Santos JG. Mechanisms of degradation of paraoxon in different ionic liquids. *J Org Chem*. 2013;78:9670–9676.

111. Pavez P, Millan D, Morales J, Rojas M, Cespedes D, Santos JG. Reaction mechanisms in ionic liquids: the kinetics and mechanism of the reaction of *O,O*-diethyl (2,4-dinitrophenyl) phosphate triester with secondary alicyclic amines. *Org Biomol Chem*. 2016;14:1421–1427.

112. MacFarlane DR, Chong AL, Forsyth M, et al. New dimensions in salt-solvent mixtures: a 4th evolution of ionic liquids. *Faraday Discuss*. 2018;206:9–28.

113. George A, Brandt A, Tran K, et al. Design of low-cost ionic liquids for lignocellulosic biomass pretreatment. *Green Chem*. 2015;17:1728–1734.

CHAPTER THREE

Polymer Mechanochemistry: A New Frontier for Physical Organic Chemistry

Luke Anderson, Roman Boulatov[1]
Department of Chemistry, University of Liverpool, Liverpool, United Kingdom
[1]Corresponding author: e-mail address: r.boulatov@liverpool.ac.uk

Contents

Abstract

Polymer mechanochemistry aims at understanding and exploiting the unique chemistry that is possible when stretching macromolecular chains beyond their strain-free contour lengths. This happens when chains are subject to a mechanical load, in bulk, in solution, at interfaces or as single molecules in air. Simple polymers such as polystyrene or polymethacrylate fragment via homolysis of a backbone C—C bond, and much contemporary effort in polymer mechanochemistry has focused on creating polymers which undergo more complex and interesting reactions, with such productive mechanochemical responses including mechanochromism and load strengthening. Comparatively less progress has been achieved in creating an internally coherent, theoretically sound interpretational framework to organize, systematize, and generalize the existing manifestations of polymer mechanochemistry and to guide the design of new mechanochemical systems. The experimental, computational, and conceptual tools of physical organic chemistry appear particularly well suited to achieve this goal, benefiting both fields.

1. INTRODUCTION

Mechanochemistry refers to a wide range of phenomena in which the kinetic stability of a molecule is affected by macroscopic motion, without changes in local temperature or pressure. Macroscopic motion that stretches macromolecular chains can dramatically accelerate reactions of their monomers, for example reducing the half-life of a covalent bond from many times greater than the age of the universe itself to the microsecond timescale at room temperature.[1]

The high anisotropy of macromolecules means that stretching them strains their constituent monomers in ways not possible in small molecules, both in terms of the magnitude of structural distortions, and its patterns. Stretching a chain of polyethylene until its half-life toward fragmentation (by homolysis of one of its backbone C—C bonds) reduces to $\sim$15 years at 300 K (corresponding to $\Delta G^{\ddagger}$ of 30 kcal/mol) increases its strain energy by 5 kcal/(mol CH_2). Both the kinetic stability and the strain energy "density" of such a stretched chain are comparable to those of hexamethyl dewar benzene and hexamethyl prismanes, whose strain energies relative to hexamethylbenzene are $\sim$60 and $\sim$120 kcal/mol, respectively, and $\Delta H^{\ddagger}$ for isomerization to hexamethylbenzene are 32 and 34 kcal/mol.[2] The importance of these strained isomers of benzene to our current understanding of chemical reactivity is well acknowledged.[3] Unlike the modest number and limited structural diversity of the highly strained molecules prepared to date, a variety of small reactive moieties can be (and have been) decorated with a pair of polymer chains. Stretching such polymers is a nearly universal strategy of straining the embedded reactive site as much as or more than any small molecule synthesized to date. Also, in contrast to the heroic synthetic effort required to obtain prismane,[4] a polymer chain can be stretched simply by subjecting its solution to ultrasound. Likewise, these polymer chains can be attached to different positions of a small reactive moiety: stretching such macromolecular "connection" isomers strains the reactive site along different molecular axes,[5,6] introducing the concept of anisotropy in discussion of reaction kinetics.

In other words, polymer mechanochemistry provides a unique means of studying the effect of extreme molecular strain on chemical reactivity. Extrapolating from the impact that studies of strained small molecules have had on our understanding of chemical reactivity, careful mechanistic investigations of mechanochemical reactions in the best traditions of physical organic chemistry will greatly broaden the range of reactions sensitive to molecular strain and mechanistic pathways for relaxation of this strain, likely reveal new patterns of strain-induced reactivity and identify new potentially generalizable strategies of controlling this reactivity by manipulating molecular structure. The new models required to describe quantitatively the effect of molecular strain on chemical reactivity in polymer mechanochemistry will likely benefit any field where intractably many molecular degrees of freedom affect chemical reactivity, from more accurate descriptions of solvation effects[7] to chemical kinetics in cells.[8] More speculatively, polymer mechanochemistry may have utility in small-molecule synthesis by enabling reactions without resorting to high temperatures or pressures,[9] although the need for polymer handles and the modest selectivity of mechanochemical reactions observed to date make this proposition somewhat less certain.[1] Mechanochemistry has traditionally been used to modify properties of polymeric materials (e.g., mastication), and more elaborate modifications by mechanochemical postpolymerization modification may be practical.[10]

Mechanochemical phenomena are multiscale processes, i.e., they result from a complex sequence of events generated by correlated motions on the length scales from ~1 μm to <1 nm and the timescale from ~1 ms to a few picoseconds.[11,12] In contrast, the vast majority of conventional chemical reactions involve correlated atomic motion only within a volume of ~1 nm^3 that is complete within a few picoseconds to 100 ns.[13] Consequently, mechanochemistry allows the conceptual and technical tools of chemistry to be applied to problems that far exceed the conventional bounds of chemistry, where "quantum meets classical and molecular meets bulk material."[14]

An equally compelling reason to study polymer mechanochemistry is its technological importance.[1,15,16] Polymers are critical to the everyday functioning of our society and subject to mechanical loads throughout their lifecycles.[17] Increasing amounts of empirical evidence suggest that the mechanochemical responses of existing polymeric materials are important (if poorly understood) determinants of their application niches. Exploiting mechanochemistry to design materials with desired molecular responses to mechanical loads is likely to lead not only to considerable improvement in existing processes and devices but also to yield fundamentally new technological solutions. For example, mechanochromic materials that change local optical properties in response to either instantaneous or cumulative loads hold potential to signal regions of a material which are most likely to fail,[18–22] a property that would be useful at every stage of material development and use. Materials which respond to localized loads that exceed a predefined threshold by forming new load-bearing bonds, thus increasing the density of bond over which the load distributes, could prevent the catastrophic failure of materials.[23–26] Combining these two productive responses to mechanical loads in a single material may offer even more exciting opportunities.[27,28]

Mechanical stress plays a key factor in many physiological processes, which offer diverse examples of (mostly) noncovalent polymer mechanochemistry. The ~1 Å elongation of the pyrophosphate P—O bond during catalytic hydrolysis of adenosine triphosphate (ATP) is transduced and amplified by kinesin and dynein motor proteins to transport objects orders of magnitudes larger than themselves as they move along microtubule tracks.[29] The unusually high toughness of structural protein titin, which is responsible for the passive elasticity of muscle, results from reversible dissociations of thousands of H bonds when the protein is stretched.[30] The phenomenology of biological mechanotransduction and the interpretational framework used to discuss its molecular basis is sufficiently distinct from those of polymer mechanochemistry to make its inclusion in the current review untenable. Fortunately, a number of thoughtful treatments of the topic have recently appeared in the literature.[31]

Advances in polymer mechanochemistry are reviewed regularly, with particular emphasis on phenomenology. Key reviews published by the end of 2016 are listed in Ref.[1] and organized by review type. We are aware of reviews that have appeared since, both phenomenological[28,32]: the objective of this review is to raise the awareness of the field of polymer mechanochemistry among physical organic chemists and to illustrate the rich research opportunities that await anyone interested in applying the formalism of physical organic chemistry to problems in contemporary polymer mechanochemistry. We are motivated by a conviction that for polymer mechanochemistry to fulfill its fundamental and technological potential, the recent explosion of empirical observations has to be matched by equally impressive advances in conceptual foundation for interpreting and generalizing experimental (and computational) findings and for enabling prediction. As a result, we focus on the aspects of polymer mechanochemistry that are rarely reviewed, especially the quantitative models currently used to discuss mechanochemical kinetics and the most common experimental methods to study mechanochemical phenomena, with particular emphasis on their limitations and outstanding unresolved questions of interpretation. Empirical observations are described only to illustrate these broader points.

2. A QUANTITATIVE MODEL OF MECHANOCHEMICAL KINETICS

The simplest example of coupling of chemical kinetics to macroscopic motion, and thus the simplest example of a mechanochemical system, is two macroscopic beads bridged by a single macromolecule moving away from each other at a constant velocity. This scenario is approximated in single-molecule force (SMF) experiments, in which a polymer chain is attached at one end to an atomic force microscopy (AFM) tip, and at the other end to a modified surface which is then retracted from the tip (see Section 3.1). If the separation of the beads is below a critical value (which depends on the chain contour length and chemical composition), the motion of the beads is suitably described by Newtonian mechanics and balance of forces (e.g., using the Langevin equation to account for thermal fluctuations). In this regime, the macromolecular bridge behaves simply as a collection of mechanical elements with a conceptually simple (if mathematically complex) relationship between extension (end–to–end distance) and the restoring force. As in a macroscopic spring, the two parameters increase simultaneously, requiring ever increasing force applied to the beads to maintain a constant retraction velocity.

Once the bead separation exceeds the critical value (typically larger than the contour length of the free macromolecule), the evolution of the system no longer follows Newtonian mechanics. In this overstretched geometry, the macrochain is chemically unstable on the timescale of the experiment, i.e., its composition, and hence the contour length, changes faster than the position of the beads. The system no longer behaves as a classical object, but rather as a quantum mechanical one. In this regime, a quantitative description of the evolution of the system requires a quantitative relationship between changes in the relative position of the beads (or equivalently, the end–to–end separation of the macromolecule) and the probabilities of the monomers making up the overstretched macromolecule undergoing chemical reaction. Such a relationship is not available within conventional chemical kinetics but must be derived by integrating the formalism of activated escape from an energy well with the variables of classical mechanics.

In theory, quantum mechanics offers a detailed description of the evolution of the system regardless of how much the macromolecular bridge is stretched. In practice, the total size of the dynamic system and hence the total number of parameters needed to describe its evolution quantum mechanically make such an approach both untenable and unlikely to yield generalizable insights. If the reactions induced by stretching the chain are highly localized, as is true for the vast majority of mechanochemistry of synthetic polymers, only a small fragment of the system described above around the reactive site where the chemical bonding changes needs to be treated quantum mechanically. It should be possible, just as it is in many other chemical problems, to coarse grain the remaining degrees of freedom of the system, both molecular (the rest of the polymer) and macroscopic (translating beads), i.e., to represent their effect on the kinetic and thermodynamic stability of the reactive site by a small number of parameters. This is true even if the macromolecule has multiple equivalent reactive sites (e.g., scission of an overstretched polyethylene chain, in which every backbone $C-C$ bond has approximately equal probability to homolyze) as long as they behave independently, i.e., do not manifest cooperativity.

The simplest implementation of this approach is a short segment of the macromolecular chain, containing the reactive moiety, attached at its terminal atoms to a compressed harmonic spring. We will postpone until later in the review discussion of important but subtle questions such as how large the segment needs to be, how thermal and ensemble effects could be treated, and what the parameters of the spring should be (see Section 2.3). This molecule/harmonic spring construct has at least three stationary states, or configurations of atoms in which each atom experiences zero net force and the restoring force of the stretched molecular fragment is identical in magnitude and opposite in direction to the restoring force F of the compressed spring (i.e., internal mechanical equilibrium). Just as in a free molecule, these states will correspond to two energy minima (reactant and product) and the saddle point separating them (transition state). Knowing the energy of the transition state relative to the reactant, $\Delta U^{\ddagger}$, allows the rate at which the molecule will change its composition (or equivalently, the probability that it will change its composition over a fixed timeframe) to be estimated using the standard transition state theory (TST) for any spring. Because the energy of a harmonic spring is a simple function of its elongation and force constant, it is productive to analyze the total energy of each state of the molecule/spring construct as a sum of the energies of the molecular component and the spring, Eqs. (1) and (2), where l_{TS} and l_R quantify the length of the compressed spring at the transition state and reactant, respectively.

$$\Delta U^{\ddagger} = \underbrace{U_{mol}^{TS} - U_{mol}^{R}}_{\text{quantum mechanics}} + \underbrace{U_{spring}^{TS} - U_{spring}^{R}}_{\text{classical mechanics}} \tag{1}$$

$$U_{spring}^{TS} - U_{spring}^{R} = \frac{k(l_{TS} - l_0)^2}{2} - \frac{k(l_R - l_0)^2}{2}$$
$$= \underbrace{\frac{k}{2}(l_{TS} + l_R - 2l_0)}_{\approx F}(l_{TS} - l_R) \tag{2}$$

If the spring is much softer than the molecular fragment (i.e., $l_0 \gg l_{TS} \sim l_R$), the restoring force of the spring, F, will be largely insensitive to the internuclear distance across which it is coupled, allowing the parameters of the spring (k and l_0) to be subsumed into the force F acting on the molecular fragment (Eq. 3). Eq. (3) can be viewed as a master equation of mechanochemical kinetics, which both underlies most quantitative discussions of mechanochemical reactivity and explains why mechanochemistry is discussed in terms of force.

$$\Delta U^{\ddagger}(F) = U_{mol}^{TS}(F) - U_{mol}^{R}(F) - F(l_{TS}(F) - l_R(F)) \tag{3}$$

Eq. (3) is valid as long as the assumptions of the TST are applicable. If force changes so fast that the strained molecule is no longer in thermal equilibrium with its environment, as may be the case in some steered molecular dynamics simulations,[33] Eq. (3) will fail, but such ultrahigh loading rates may not be experimentally accessible, making such failure of no practical significance.

Practical applications of Eq. (3) require knowledge of U, $l_R(F)$, and $l_{TS}(F)$. For most (but not all) localized chemical reactions, these parameters are available with a varying degree of accuracy from quantum chemical calculations.[1] An important exception is mechanochemiluminescence of 1,2–dioxetanes,[21,26] whose complex nonadiabatic dissociation mechanism[34,35] is not yet amenable to usefully accurate quantum chemical calculations of force-dependent barriers. Much empirical discussion of mechanochemical reactivity, however, remains based on various approximations of Eq. (3), and the plethora of reported equations of mechanochemical kinetics, from the original Eyring–Bell (EB) ansatz to the more recent two-dimensional and complete harmonic models, are simplifications of Eq. (3). Most often these approximate models are applied to measured (usually by single-molecule force spectroscopy—SMFS) force-dependent reaction kinetics to estimate a structural parameter of the transition state $(l_{TS}(0) - l_R(0))$. This parameter is also used as an empirical quantifier of how sensitive the reaction rate is to stretching of the molecule.

2.1 Approximate Solutions of the Master Equation of Mechanochemical Kinetics

2.1.1 Zeroth-Order Approximation: The EB Model

The simplest approach to evaluating Eq. (3) is to assume that stretching the molecule has no effect on either its intrinsic reactivity (i.e., $U_{TS}(F) - U_R(F) = U_{TS}(0) - U_R(0)$) or the structural differences between the reactant and transition states $(l_{TS}(F) - l_R(F) = l_{TS}(0) - l_R(0))$. These assumptions reduce Eq. (3) to

$$\Delta U^{\ddagger}(F) \approx \Delta U^{\ddagger}(0) - F(l_{TS}(0) - l_R(0)), \tag{4}$$

or the more commonly shown expression for the force-dependent rate constant, $k(F)$, given by Eq. (5), or the corresponding survival probability (i.e., the probability that the molecule will not have reacted by reaction time t), $S(t, F)$. In the literature, the $l_{TS}(0) - l_R(0)$ term is often written as $\Delta x^{\ddagger}$. k_B is the Boltzmann constant; $k(0)$ and $S_0(t)$ are the parameters in strain-free molecule.

$$k(F) = k(0)e^{\frac{F(l_{TS}(0) - l_R(0))}{k_B T}} \tag{5}$$

$$S(t, F) = [S_o(t)]^{e^{-\frac{F(l_{TS}(0) - l_R(0))}{k_B T}}} \tag{6}$$

If the force is time dependent, the zeroth-order approximation of Eq. (3) can only be expressed in general as survival probability (Eq. 7).

$$S(t, F) = e^{-k(0)\int_0^t e^{\frac{F(\tau)(l_{TS}(0) - l_R(0))}{k_B T}} dt} \tag{7}$$

The zeroth-order approximation of mechanochemical kinetics is often traced to the work of Eyring, who considered the flow of polymer chains,[36] and to a later paper by Bell, in the context of cell adhesion.[37] An important but mostly overlooked difference between the zeroth-order models in use today and those considered by Eyring or Bell is that both researchers simply postulated that barrier height depends linearly on applied force without any attempt to define the proportionality constant (apart from the obvious statement that it must have a dimension of length), much less ascribe it to specific changes in molecular geometry. At present, probably the most common misuse of the zeroth-order approximation (often referred to as the Bell or EB equation) is to determine the elongation of a scissile bond in the transition state of a reaction from SMF experiments. These attempts often use a version of the zeroth-order approximation which includes time-dependent force, first considered by Evans,[38] who derived Eq. (7) from the Kramers formulation of chemical kinetics.[39,40]

Eqs. (4)–(7) predict that a reaction whose transition state is longer than the reactant along the constrained axis $(l_{TS}(0) - l_R(0) > 0)$, is accelerated by tensile force ($F > 0$ by convention). This acceleration results solely from the decrease in the strain energy of the compressed spring, enabled by the localized elongation of the reactive site. Since the spring represents molecular degrees of freedom of the stretched macrochain removed from the reactive site, the zeroth-order approximation of mechanochemical kinetics postulates that tensile force accelerates a chemical reaction when the formation of the transition state allows partial relief of molecular strain in the nonreactive degrees of freedom. To achieve this partial strain relief, the reactive site should elongate in the transition state. The EB model remains by far the most commonly used model of mechanochemical kinetics due to its conceptual and technical simplicity, despite substantial evidence of its shortcomings.[41–43]

It is sometimes suggested that Eq. (4) can be improved by replacing the $F(l_{TS}(0) - l_R(0))$ term with a path integral along the reaction coordinate. This is wrong, however, because the TST requires the activation energy to be a state function, i.e., a function whose value only depends on the initial and final state and not on the path connecting the two (Fig. 1).

2.1.2 First-Order Approximation: Tilted Potential Energy Surface (TPES) and Cusp Models

The assumption that stretching a molecule does not change its intrinsic reactivity clearly contradicts a large body of literature attesting to the strong effects of molecular strain on chemical kinetics in nonpolymeric substrates (where the relaxation of the molecular degrees of freedom outside the reactive site contribute minimally to the changes in reaction barriers, i.e., the $F(l_{TS} - l_R)$ term of Eq. 3 is irrelevant). The first-order approximations of Eq. (3) all assume (either implicitly or explicitly) that the constrained distance is a normal mode of the reactant and either a normal mode of the product (sometimes referred to as "cusp model," Fig. 2B) or the reactive

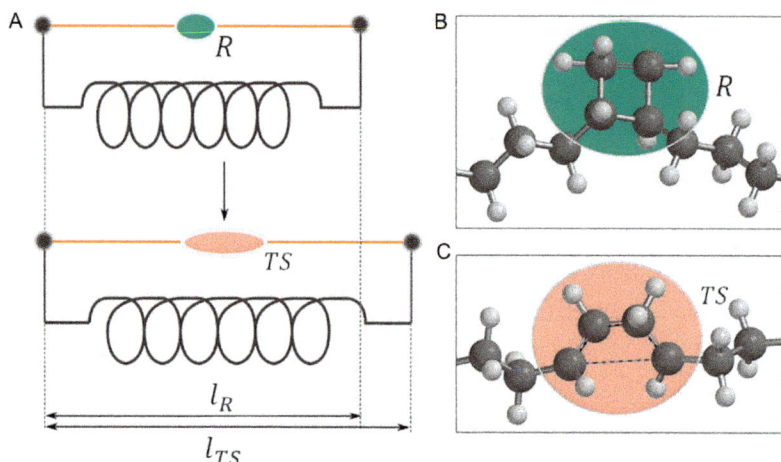

Fig. 1 (A) A reactive site (*green circle*) in a polymer fragment stretched by an attached compressed harmonic spring reacts through a transition state (*pink oval*) which is longer along the constrained axis than the reactant state. The composition of the molecular fragment outside of the reactive moiety (*orange*) is unchanged. The resulting lengthening of the compressed spring reduces its strain energy, thus lowering the reaction barrier. (B) and (C) An example of such a reactive site is cyclobutene, whose transition state for electrocyclic ring opening is ~2 Å longer than the reactant. C atoms are represented by *dark gray spheres* and H atoms by *lighter gray spheres*.

mode of the transition state (in which case the model is called "tilted potential energy surface"[44] or "extended Bell model"). Other formulations of the TPES or cusp model appear in the literature, but they all reduce to the assumption above. Only under the assumption of the constrained distance being a normal mode or the reactive mode can the force dependence of the energy of each stationary state and of the constrained distance be expressed as a function of the same single parameter of that state, k_i ($i = R$, TS, or P, Eqs. 8 and 9), whose physical meaning is the force constant of the constrained distance. Note that while k_R and $k_P > 0$, $k_{TS} < 0$. What happens with the equations (and the model) when this aphysical assumption is relaxed is discussed below.

$$U_i(F) = U_i(0) + \frac{F^2}{2k_i} \tag{8}$$

$$l_i(F) = l_i(0) + \frac{F^2}{k_i} \tag{9}$$

Substituting $U_i(F)$ and $l_i(F)$ of master Eq. (3) by Eqs. (8) and (9) yields Eq. (10) for the TPES model, which deviates from the zeroth-order (EB)

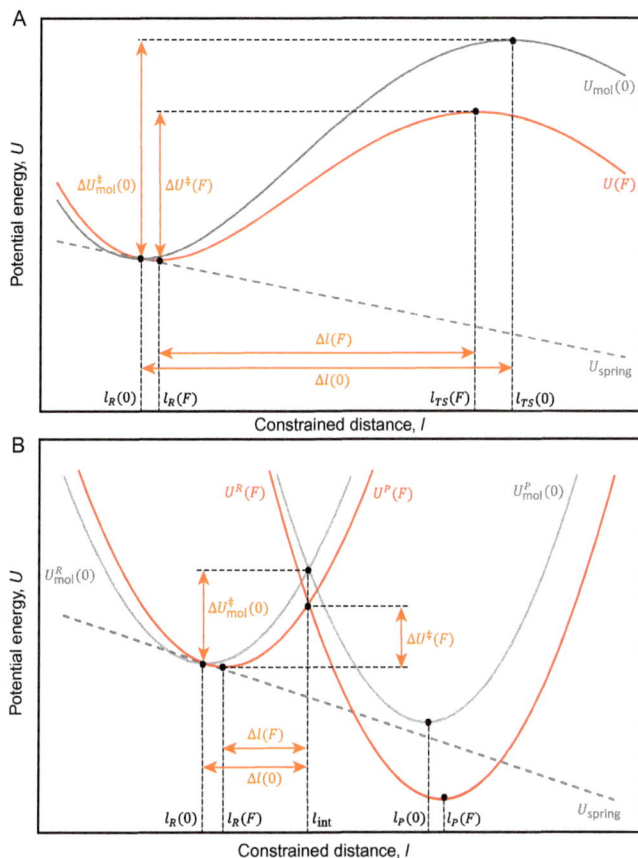

Fig. 2 Illustration of (A) the tilted potential energy surface and (B) the cusp model for the first-order approximation of the effect of external force on chemical reactivity.

model (Eq. 4) through the presence of the quadratic term $-\frac{F^2}{2}\left(\frac{1}{k_{TS}} - \frac{1}{k_R}\right)$. Because the TPES is internally consistent only if $k_{TS} < 0$, this quadratic term must be positive and increase the reaction barrier relative to the zeroth-order (EB) estimate. The outcome is a manifestation of the Hammond effect, which postulates that a barrier-lowering perturbation makes the reactant and the transition state more "alike," both structurally ($[l_{TS}(0) - l_R(0)] > [l_{TS}(F) - l_R(F)]$) and energetically ($[U_{mol}^{TS}(0) - U_{mol}^{R}(0)] > [U_{mol}^{TS}(F) - U_{mol}^{R}(F)]$).[45] While the latter lowers $\Delta U(F)$, the former decreases the strain energy of the coupled spring that is released in the transition state, because it decreases how much the compressed spring relaxes. The quadratic dependence means that difference between the zeroth- and first-order approximations increases with force. The TPES is illustrated in Fig. 2A.

$$\Delta U^{\ddagger}(F) \approx \underbrace{\Delta U^{\ddagger}(0) - F\big(l_{TS}(0) - l_R(0)\big)}_{\text{EB (zeroth-order) model}} - \frac{F^2}{2}\left(\frac{1}{k_{TS}} - \frac{1}{k_R}\right) \qquad (10)$$

The TPES operates only if the constrained distance is the reactive mode in the vicinity of the transition state, otherwise, Eqs. (8) and (9) ($i = TS$) are invalid (see next section). In practice, no multiatomic molecule has ever been reported that satisfies this criterion, and the TPES has been used to discuss force-dependent barriers of reactions in which both k_R and k_{TS} are positive. Such use is sometimes rationalized by claiming that F is a "component" of the applied force acting along the strain-free reaction coordinate. However, such a distinction is meaningless because Eqs. (8) and (9) are simultaneously valid only if the constrained distance is a normal (reactive) mode; otherwise, force will increase the molecular energy faster than proportionally to the elongation of the constrained distance. Whether the internal inconsistency between the underlying assumption of the model and the numeric parameters used within is the main reason for the aphysical kinetics predicted by Eq. (10), even when the more obvious errors are avoided, is unknown.

In the cusp model, the activation barrier height is determined by the intersection of the energy wells corresponding to the reactant and the product (Fig. 2B). In contrast to the TPES, the length of the constrained distance at the intersection, l_{int}, is independent of the stretching force, allowing the master equation to be reduced to the approximate form of Eq. (11).

$$\Delta U^{\ddagger}(F) \approx \Delta U^{\ddagger}(0) - F\big(l_{\text{int}}(0) - l_R(0)\big) + \frac{F^2}{2k_R} \qquad (11)$$

Note that the only difference between Eqs. (10) and (11) is the absence of the $F^2/2k_{TS}$ term, which is expected since in the cusp model the intersection is equivalent to an infinitely stiff transition state ($k_{TS} = \infty$). Given the similarity between Eqs. (10) and (11), especially at low forces, it is somewhat surprising that in the few cases where the same experimental data were fitted to these equations, fairly large differences in $l_{TS}(0) - l_R(0)$ were obtained.[46] The reason that the cusp model predicts a higher activation barrier than the EB model is the same as in the TPES: a Hammond effect shift of the reactant geometry toward the intersection that decreases the energy the constraining spring releases as the molecular geometry changes from the (shifted) reactant to the (stationary) transition state.

While the cusp model avoids the aphysical assumption of the constrained distance being the reactive mode in the vicinity of the transition state, it is only applicable for nonscissile mechanochemical reactions, i.e., reactions in which the macromolecule does not fragment. Otherwise, the assumption of an infinitely compliant constraining spring (to achieve the constant stretching force irrespective of the structural differences between the reactant and the transition state or intersection) would correspond to an infinite reaction energy and an infinite elongation of the constrained internuclear distance in the product.

The cusp model is equivalent to the Marcus theory of electron transfer.[47,48] While the Marcus-like treatment of nucleophilic displacement reactions was previously shown to be a productive strategy in rationalizing structure/reactivity relationships in the gas phase,[49] it is not obviously conceptually acceptable to treat an arbitrary reaction in a stretched polymer as a nonadiabatic process. Practically, the other assumptions of the model, particularly the validity of Eqs. (8) and (9), may introduce considerably greater errors in the derived parameters than the assumption of nonadiabaticity or an infinitely rigid transition state.

2.1.3 Second-Order and Complete Harmonic Approximations

The obviously aphysical assumption of the first-order models (so-called TPES, extended Bell or cusp models) prompted some attempt to analyze force-dependent reaction kinetics in terms of two-dimensional reaction surfaces, with one coordinate being the constrained internuclear distance l, and the other a collective coordinate that represents the remaining 3N-7 nuclear degrees of freedom, q.[50] This approach again follows the well-established precedent of physical organic chemistry (e.g., the "Bema–Hapothle" model[51]) and suffers from the same limitations, as discussed elsewhere.[12] Within this model, the expressions for the molecular energy of state i (R or TS) is given by Eq. (12), where k and k' are harmonic force constants of the constrained distance l and the collective coordinate q, and k'' is the coupling constant (equivalent to the "interaction parameter" of the Bema–Hapothle model). The coupling constant k'' is a direct consequence of the constrained distance not being a normal mode of the molecule and hence l and q not being orthogonal. Otherwise, $k'' = 0$ and Eq. (12) reduces to Eq. (10) (first-order models). Note that unlike k_{TS} of the first-order model, the second-order model makes no assumption of which of the three force constants of the TS are negative, and the coupling constant k''_R can be either negative or positive.

$$U_i(F) = U_i(0) + \frac{F^2}{2\left(k_i - \frac{k_i^{''}2}{k_i'}\right)} \qquad (12)$$

The consequence of nonzero coupling between the two coordinates of the second-order approximation is that constraining a single internuclear distance of a molecule to a nonequilibrium value can cause other internuclear distances to elongate or contract, even if those distances are orthogonal to the constrained distance in 3D Cartesian (physical) space (Fig. 3). Likewise, whereas the first-order models predict that tensile force acting on the transition state always lowers its energy ($U_{TS}(F) < U_{TS}(0)$), force

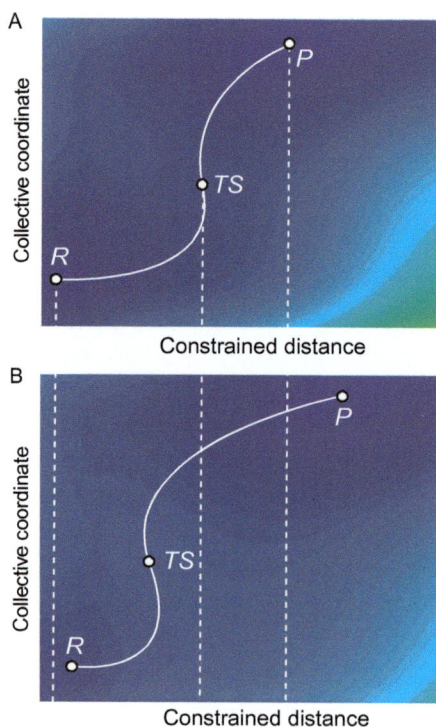

Fig. 3 A two-dimensional potential energy surface for a reaction occurring in reactive moiety within a macromolecule (A) in the absence of force and (B) in the presence of a compressed spring, which qualitatively changes the position of the reactant, transition state, and product, and the minimum energy reaction pathway (*all highlighted*). Note that the reactant and transition state do not necessarily become closer along the constrained distance upon stretching the macromolecule. *Warmer and colder colors* correspond to higher and lower energies, respectively.

can lower or raise the energy of the transition state, depending on the relative magnitudes of the three harmonic constants, k_{TS}, k'_{TS}, and k''_{TS}. In that respect, the second-order model is a considerable improvement over the first-order models, because every quantum chemical calculation reported to date revealed that force destabilizes transition states even when $\Delta U(F)$ $< \Delta U(0)$, due to partial relaxation of the coupled spring.

The corresponding approximation of the master equation is given by Eq. (13), which, as expected, reduces to Eq. (11) if l is assumed to be a normal/reactive mode (i.e., $k''_R = k''_{TS} = 0$).

$$\Delta U^{\ddagger}(F) \approx \underbrace{\Delta U^{\ddagger}(0) - F\big(l_{TS}(0) - l_R(0)\big)}_{EB \text{ (zeroth-order) model}} - \frac{F^2}{2}\left(\frac{1}{k_{TS} - \dfrac{k''^2_{TS}}{k'_{TS}}} - \frac{1}{k_R - \dfrac{k''^2_R}{k'_R}}\right)$$

$$(13)$$

The main difference between the second-order and first-order models is that the former accommodates anti-Hammond effects, and $\Delta U(F)$ estimated by Eq. (13) can be larger or smaller than $\Delta U(F)$ estimated by the EB model (an internally consistent first-order approximation of $\Delta U(F)$ always exceeds the EB estimate of $\Delta U(F)$, although the reported applications of Eq. 11 are not internally consistent). In other words, the second-order correction to the activation energy will in general be nonzero even if the constrained coordinate l is orthogonal to the reaction path in the vicinity of R, TS, or both (Fig. 3). The practical significance of this additional flexibility of Eq. (13) is not clear, because at present there does not appear to be any theoretically valid method of estimating k' and k'', e.g., by quantum chemical calculations, precluding prediction of $\Delta U(F)$ from strain-free molecular parameters using Eq. (13). Conversely, any coefficients derived from fitting experimental data to Eq. (13) would lack a clear molecular interpretation.

It is conceptually trivial to extend the second-order approximation to a nonredundant set of 3N-6 internal coordinates in the harmonic approximation.[43,52] If the first coordinate of the set is the constrained distance, this full harmonic approximation of master Eq. (3) is given by Eq. (14), where $C_i(1,1)$ is the compliance of the constrained coordinate in state i. The compliance is obtained by inverting the full Hessian matrix (a matrix of the second-order derivatives of energy with respect to the 3N-6 internal coordinates) of the strain-free geometry. Most quantum chemical methods produce this Hessian in an analytical frequency calculation, but the large size of a

Hessian of even a moderately sized substrate makes accurate calculations of its inverse technically challenging. In our own work,[41,42,53–61] we found that molecular compliances of molecules containing up to 100 atoms can be calculated with useful accuracy on geometries converged to RMS force of $<10^{-6}$ a.u., using high accuracy integration grids and Cholesky-decomposition method for matrix inversion.

$$\Delta U^{\ddagger}(F) \approx \underbrace{\Delta U^{\ddagger}(0) - F(l_{TS}(0) - l_R(0))}_{EB\,(\text{zeroth-order})\,\text{model}} - \frac{F^2}{2}(C_{TS}(1,1) - C_R(1,1)) \quad (14)$$

Partial compliance matrices of only a few reactant/transition state pairs have been reported. In all cases, $C_{TS}(1,1) > 0$ but no generalizable relationships between C_{TS} and C_R has emerged, so that the harmonic approximation estimate of $\Delta U(F)$ can be smaller or larger that the EB estimate. In other words, the constrained distance in some transition states is softer than in the reactant $(C_{TS} > C_R)$, while in others the reverse is true. The advantage of Eq. (14) is that all elements have rigorously defined molecular interpretations and are available from quantum chemical calculations. The approach underlying Eq. (14) also allows the restoring force of any molecular degree of freedom (e.g., an internuclear distance) in addition to that of the explicitly constrained distance to be calculated as a function of F, which is of practical and conceptual importance. In general, Eq. (14) allows much more accurate predictions of force-dependent barriers by replacing the constrained distance l and the applied force F with a properly selected internuclear distance and its restoring force (see the next section).

2.2 Accuracy of the Conventional Approximations and Systematic Strategies of Improving Them

The very few reported analyses of the accuracy of various approximation of Eq. (3) were performed by comparing the force-dependent activation free energies estimated by Eqs. (4), (10), (11), (13), or (14) to $\Delta U(F)$ values calculated quantum chemically.[41,43,60] Most of these analyses did not consider the question of how faithfully $\Delta U(F)$ values from quantum chemical calculations reproduced physical reality, which is a complex and largely unresolved problem in polymer mechanochemistry, as discussed elsewhere.[1] Invariably, the zeroth-order approximation of Eq. (3) produced the largest errors and the "complete" harmonic approximation (Eq. 14) performed the best, with the accuracy decreasing at larger forces. The main cause of the (often substantial) errors is anharmonicity of the constrained distance, which is equally problematic for unimolecular and bimolecular reactions. Anharmonicity increases with the size of the molecular moiety (i.e., the macromolecular segment that is treated atomistically instead of being modeled as a compressed harmonic spring) and cannot be eliminated simply by making the molecular fragment smaller (which makes the constrained coordinate stiffer and generally more harmonic), as it introduces its own artifacts.[1,42,60] The so-called Taylor expansion and local coordinate approaches considerably improve the estimates of $\Delta U(F)$ at minimal incremental computational cost.

The Taylor expansion approach recognizes that all approximate solutions of Eq. (3) (i.e., Eqs. 4, 10, 11, 13, or 14) are Taylor expansions of $\Delta U(F)$ with respect to F truncated as the first or second term. It was therefore suggested[62] that as long as the approximations of Eq. (3) are used to empirically quantify the reaction "sensitivity" to stretching force from experimental measurements (e.g., SMF experiments) a more productive approach may be to fit the experimental data to a Taylor expansion of $\Delta U(F)$ truncated at the highest order compatible with the quality and quantity of the experimental data. In this approach, the fitting parameters would be analogous to nucleophilic constants of physical organic chemistry that could be usefully compared across different reactions, polymer architectures, and/or loading conditions. The measured force-dependent kinetics reported to date appears to lack the accuracy and/or dynamic range needed to estimate the Taylor coefficients of second- or higher order with useful accuracy, but such an analysis may be more productive than attempting to ascribe any deviation of the measured correlations from those predicted by EB to changes in the underlying reaction mechanisms (which are difficult if not impossible to extract from SMF experiments alone). The lack of molecular interpretation of the Taylor expansion coefficients beyond the second-order means that at present the Taylor expansion approach does not allow estimates of $\Delta U(F)$ to be improved systematically from computed parameters of strain-free reactants and transition states.

The local coordinate approximation appears to be the most general and promising approach to dealing with anharmonicity of the constrained distance.[41,42,53–56,58,59,61] Stretching a molecule by constraining one of its non-bonding distances to a nonequilibrium value distorts other molecular coordinates (distances, bond angles, and torsions), and the magnitude of this distortion is captured by the restoring force of each coordinate. The molecular compliance matrix C (e.g., Eq. 14) allows the restoring force of any internal coordinate to be expressed as a function of the stretching force F.

Each reactive site studied to date appears to have at least one internuclear distance (a) that is considerably more harmonic (i.e., its restoring force is approximately proportional to its absolute strain) than the constrained distance and (b) whose restoring force, F, is a much better predictor of the activation energy of the stretched reactant than the applied force F using Eq. (13). Furthermore, for reactions of the same mechanistic type (e.g., S_N2 displacements), force-dependent activation barriers of multiple structurally-distinct reactants is predicted accurately using the same internal coordinate, the separation of the two atoms that connect the electrophilic atom to the polymer.[41,42,54,56,57,59–61]

The local coordinate approximation improves the accuracy of predicting the reaction kinetics in a stretched macromolecule in the harmonic approximation (i.e., using Eq. 14).[41,42,60] It also simplifies such predictions technically by allow them to be separated into two simpler problems: estimating the activation barrier as a function of the restoring force of a local coordinate in a minimal reactant and estimating the relationship between the applied force and this local restoring force. A rigorous definition of the minimum reactant does not exist and appears to depend on the reaction: dimethylcyclobutene,[41] dimethyldibromocyclopropane,[58] and tetramethylpyrophosphate[57] were all demonstrated to yield local force/activation energy correlations that accurately extrapolate to larger homologs, including polymers. The use of the minimal reactant increases the maximum level of theory at which the strain-free geometries and energies can be practically calculated and decreases the number of conformers that need to be optimized for correct calculations of free energies (see next section). The relationship is independent of the polymer in which the reaction occurs and hence applicable to polymers with different backbones by combining it with the dependence of the local restoring force on the applied force.

This latter parameter, sometimes called the chemomechanical coupling coefficient, defines the capacity of the polymer backbone to transmit the applied force to these sites or redistribute it away from them. The chemomechanical coupling coefficient of a few simple backbones (e.g., polyesters, acrylates, and simple aliphatic hydrocarbons designed to model polystyrene) was reported.[41] Importantly, the calculated values appear to be rather insensitive to the model chemistry, with even semiempirical methods (e.g., PM6) giving acceptable results, and values which are similar in both the reactant and the transition state. The latter comparison, performed at the DFT level, seems to suggest that the coupling coefficient is determined primarily by the micromechanics of the polymer backbone, which is insensitive to the bonding pattern of the reactive site, rather than by vibrational coupling between the degrees of freedom of the two fragments. The chemomechanical coupling coefficient depends on the length of shorter polymer segments but reaches a length-independent value for a few repeat units.

By explicitly defining the two independent determinants of reaction kinetics in stretched polymers (the intrinsic mechanochemical reactivity of the localized sites and the chemomechanical coupling coefficient), the local coordinate approximation enables practical calculations of single-chain micromechanics of mechanochemically labile polymers,[25,58,63] simplifies identification of broad structure/activity correlations,[62] and facilitates the design of polymers with predetermined mechanochemical profiles by enabling independent control over the chemistry and the critical force above which the chemistry is observed.[64,65]

The observation that the local restoring force enables accurate predictions of localized reaction kinetics in stretched polymers[58] establishes a conceptual connection between molecular strain and chemical reactivity irrespective of the size of the reactant, or the manner in which strain is imposed on the reactive site: by incorporating it into a strained nonpolymeric molecule or by stretching a macrochain containing the site in its backbone. The only condition is that the imposed strain is anisotropic. In macromolecules, this anisotropy is imposed by the very large aspect ratios of the polymers; in small molecules, only specific molecular architectures are expected to achieve comparable degree of anisotropy[62] (which can be estimated with the help of the compliance matrix).

2.3 Complicating Factors: Ensemble Effects, the Minimum Length of the Macromolecular Segment, Multibarrier Reactions, and Competing Mechanisms

Up to this point, the discussion of the quantitative model of mechanochemical kinetics (i.e., Eq. 3) has omitted any consideration of how the model predictions are affected by the length of the atomistically treated polymer segment, or the fact that the reactant and transition states of molecules of interest in polymer mechanochemistry are conformational ensembles, i.e., comprised of multiple conformers in rapid equilibrium. Quantum chemical calculations, performed on homologous series of diverse reactants, including cyclobutenes, cyclopropanes, phosphotriesters, and siloxanes, suggest that both effects are significant, particularly at forces <2 nN.[12,42,56,57,59,60]

For bimolecular reactions, such as hydrolysis of pyrophosphate esters, approximating each state by its lowest energy conformer systematically overestimates calculated $\Delta U(F)$ by up to 3 kcal/mol by neglecting so-called conformational entropy effects.[57] In these reactions, the higher coordination number of the electrophilic atom in the transition state relative to the reactant means that the transition state is comprised of significantly fewer thermally accessible conformers (i.e., conformers within ~1.5 kcal/mol of the conformational minimum) than the reactant. Equivalently, the reactant state is enriched in conformers with particularly short end-to-end separations (across which force is applied). These additional reactant conformers are particularly strongly destabilized by tensile force, reducing the total number of thermally accessible conformers in the reactant state and increasing the free energy of the reactant state relative to that of the minimum energy conformer of the reactant state faster with tensile force than the free energy of the transition state relative to that of its minimum energy conformer. In other words, the existence of a conformational ensemble leads to greater destabilization of the reactant state by force, and hence, a larger decrease in the activation free energy, that would be predicted by considering only one conformer for each state.

In extreme cases, ignoring the fact that rates are governed by relative energies of states rather than individual conformers leads to qualitatively incorrect predictions. For example, neutral methanolysis of siloxanes is calculated to proceed by two competing two-step mechanisms, one of which is accelerated by force and the other inhibited. Considering only the minimum energy conformers predicts the two mechanisms to be approximately iso-energetic, whereas the force-inhibited path has a lower strain-free activation free energy when complete conformational ensembles are used. The difference is larger for larger homologs and no size-independent limit of force-dependent activation energy for this reaction was identified.[42]

Eq. (3) and its complete harmonic approximation (Eq. 14) were extended to free energies using the statistical mechanics formalism.[52] Note that while the approximation of the infinitely compliant spring underlying Eq. (3) is simply a convenience and potential energy of activation can be calculated for a coupled spring of any compliance (irrespectively of whether the result is physically relevant[1]), a closed-form expression of activation free energies could only be obtained if the coupled spring was infinitely compliant.

The application of Eq. (3), regardless of how the terms are estimated (i.e., from explicit quantum chemical calculations at multiple forces or by extrapolating strain-free parameters) to unimolecular (single barrier) reactions is conceptually straightforward. Estimates of force-dependent kinetics of multibarrier reactions require more care because relative energies of individual stationary states (i.e., intermediate and transition states) in general manifest different dependencies on force, often leading to changes in the rate-determining step.[25,42,61,63] Additionally, it is not uncommon for force to stabilize intermediates below the reactant, in which case the intermediate may accumulate and the overall reaction rate may be determined by the rate of decay of this intermediate rather than that of the reactant.[25] In such cases, the dependence of the reaction rate on force can change significantly as the force increases. Failure to account for such changes, for example, by equating the total activation energy to the extrapolated energy of the rate-determining transition state in a strain-free molecule can lead to qualitatively incorrect predictions.[1]

The role of competing reaction pathways in determining mechanochemical reactivity was ignored until recently.[25,42,63,66] It is very likely that for most if not all mechanochemical reactions studied to date multiple reaction mechanisms are kinetically competitive over a range of applied forces. This is particularly true if the minimum energy path in strain-free reactant is inhibited by force. Known examples include isomerizations of *cis*-dimethylcyclobutene and its derivative,[46] and of *trans*-dimethyldihalocyclopropanes[67] and retro–Diels–Alder reactions of certain adducts of anthracene.[1,6] In all cases, the minimum energy concerted reaction mechanisms in strain-free reactants are strongly destabilized by force whereas the higher-energy diradical alternative is stabilized by force. Neglecting the existence of such competing paths may lead to qualitatively incorrect predictions of mechanochemical reactivity, as illustrated recently.[1,6] In the absence of robust automated procedures for finding all reaction pathways that a molecule can follow, including those that may seem kinetically inconsequential in strain-free reactants much depends on one's breadth of empirical knowledge of chemical reactivity.

3. EXPERIMENTAL TECHNIQUES OF POLYMER MECHANOCHEMISTRY

Macromolecules are easy to stretch, but hard to maintain at a well-defined accurately known strain (or equivalently, restoring force). Twisting, stretching, or compressing a macroscopic sample of a polymer stretches some fraction of polymer segments. This method, while simple, allows no control over the magnitude of the distortion of individual chains or how long the fragments are maintained in the stretched state. At the other extreme is SMFS, which allows segments of individual isolated macromolecular chains to be stretched until their fragmentation at rates from $<10 \, nm/s$ to $> 10 \, \mu m/s$

or maintained at approximately constant strain or restoring force for up to a few seconds. The cost of this control is the technical complexity, with only a handful of groups worldwide able to perform SMF experiments relevant to polymer mechanochemistry. In between these two extremes are several techniques of vastly different technical difficulties for stretching macromolecules, with varying degree of control over the imposed strain. Many of these techniques rely on coupling between a macromolecular solute and hydrodynamic flows to stretch chains.

3.1 SMFS

A macrochain or its segment can be stretched and maintained in a stretched conformation only if it couples to its surroundings at a minimum of two atoms. To a useful approximation, this conceptually simplest mechanism of chain stretching is realized in SMFS. In theory, in an SMF experiment a polymer chain is anchored to a functionalized surface at one end, and to a microcantilever tip on the other (in practice, parts of the chain are often absorbed at one or both surfaces, an effect which is discussed in greater detail later in this section). The surface is then retracted from the tip, extending the end-to-end separation beyond its equilibrium value, thus stretching the chain. The restoring force of this stretched chain deflects the cantilever, with this deflection measured and converted to the force using a number of empirical equations.[68] Note that SMF experiments do not directly measure the restoring force, as is often mistakenly assumed.

SMF experiments are performed in one of two modes: constant velocity (dynamic force spectroscopy) and constant force (force-clamp). In the former, the two surfaces anchoring the macromolecular bridge are retracted at a constant velocity and the data are recorded as a force/extension curve. In the latter, the chain is rapidly stretched to a desired restoring force and then maintained at this force by moving the surface: the extension is recorded as a function of the time the chain is maintained at a fixed force.

In dynamic force spectroscopy, the strain experienced by the chain increases continuously until the half-life of one or more of its monomers decreases to the milliseconds timescale, at which point a localized reaction happens. If the reaction either breaks the chain or increases its strain-free contour length by at least ~ 1 nm, a mechanical instability, in which the extension increases while the force decreases, becomes resolved from the usual thermal fluctuations. This mechanical instability is a direct consequence of the vastly different timescales required for local rearrangement of chemical bonding that constitutes a reaction and reestablishment of mechanical equilibrium between a stretched macrochain and AFM tip.[52] The abrupt rearrangement of local bonding occurs on the 10–100 ps timescale, followed by a redistribution of macromolecular conformers on the ~1 μs timescale (corresponding to the longest relaxation time of a macrochain with contour length on the order of 1 μm) to accommodate the new local geometry. On these timescales, the AFM tip is stationary (e.g., its thermal fluctuations occur on <1 kHz scale) and therefore out of mechanical equilibrium with the suddenly elongated chain. This longer chain allows partial relaxation of the bent AFM tip, which is recorded as a decrease in the applied stretching force. The larger the difference in the contour length of the polymer before and after the reaction, the larger this force drop is. If a stretched macromolecule is made of multiple mechanochemically reactive nonscissile sites, multiple mechanical instabilities are resolved, producing a "sawtooth" pattern. If an individual reaction increases the polymer contour length by <1 nm, multiple sites will produce a plateau on the force/extension curve where the extension increases at approximately constant force, instead of the sawtooth pattern. Continued stretching of the chain eventually results in the failure of the macromolecular bridge by any number of plausible reactions.

The force-clamp mode of SMFS relies on a feedback mechanism, whereby the position of the glass slide relative to the AFM tip is varied at below kiloHertz frequency to maintain the constant deflection of the AFM tip (and hence an approximately constant stretching force acting on the polymer). The data are recorded as extension vs time. The extension of a chemically inert macromolecule would fluctuate randomly within an Å-scale range, determined by the stiffness of the cantilever and the temperature. A reaction occurring within the stretched macromolecule that increases its contour length results in an abrupt increase in the recorded elongation. Multiple reactions will produce a "staircase" extension profile, equivalent to the sawtooth pattern observed in force-ramp SMFS. While performing SMF experiments in the force-clamp mode may seem appealing to avoid the complications of time-dependent rate constants (e.g., Eq. 7), technical limitations, including the narrow range of accessible forces, poorer resolution of extension, and thermal drift restrict the application of the force-clamp mode in polymer mechanochemistry. A recently published study comparing the results obtained by the two methods suggested that they yield approximately equivalent information.[69]

Molecular interpretation of SMF experiments, either qualitative (which reaction occurred, and by what mechanism) or quantitative (at what rate the reaction proceeds, and how this rate depends on force), is far more challenging than is generally acknowledged by the practitioners of the field and could definitely benefit from a closer cooperation with physical organic chemists than have been the case thus far. SMF experiments remain the only experimental technique of estimating mechanochemical kinetics of diverse reactions and are therefore critical for advancing our understanding of chemical reactivity in highly stretched macromolecules, for improving and validating new experimental and computational methods of quantifying macromolecular reactivity, and for developing applications of mechanochemical phenomena. Improvements in the quality of data that are available from SMF experiments are likely to have a disproportionate impact on the development of polymer mechanochemistry as a bona fide discipline. However, such improvements require broad awareness of and an agreement on the primary determinants of the reliability of molecular interpretations of SMF experiments. It is with this objective that we articulate below our understanding of these determinants.

At present, SMF experiments do not allow spectroscopic characterization of the reaction products. Signatures of mechanical instability, whether resolved (sawtooth) or not (plateaus) in dynamic SMFS, or abrupt increases in the chain extension in force-clamp SMFS are the only indication that a reaction has occurred in these experiments. Reactions that do not increase the polymer contour length are thus invisible to SMFS. The nature and localization of scissile reactions that fragment the macromolecular bridge can rarely be identified reliably, and such reactions should be viewed as largely unsuitable for SMF studies[42] (an important exception is SMFS of polymers containing a single labile backbone bond, e.g., Ref.[70]). In the other extreme, resolving contour length elongations resulting from reactions of individual nonscissile reactive sites will most likely yield credible identification of the underlying chemistry if the observed distribution of individual chain elongations matches that obtained by high-quality quantum chemical calculations of the assumed reaction using a fragment of the stretched macromolecule. A nonscissile reactive site reacts mechanochemically with an increase of the polymer contour length instead of chain fragmentation. So far only one reported study,[25] mechanochemical dissociation of cinnamate dimers, has achieved such resolution, but the molecular design used to ensure the sufficiently large increase in the contour length upon dissociation of individual dimers is general enough to be applied to many other reactions.

Significantly more numerous are examples of SMF experiments on polymers containing multiple equivalent nonscissile reactive sites, where reactions of individual sites cannot be resolved in the force/extension curves, because each increases the contour length only by a few Å.[32] Such small increments produce a plateau in force/extension curves without a sawtooth pattern and the only reliable approach reported to date of verifying the nature of the reaction is to compare the full experimental force/extension curve, including the regions before and after the plateau, to the curve extrapolated from quantum chemical calculations of the candidate reaction(s).[58,63] The need to model the portions of the force/extension curve where no reaction happens (either because the reactive sites are too stable kinetically at low force or because all reactive sites have already reacted at high force) is to estimate the number of the reactive sites independently of how much each elongates (conversely, the length of the plateau is a product of these two unknowns). Short of such modeling, no reliable means exist of estimating the number of reactive sites per strain-free contour length of the macromolecule with useful accuracy from the force/extension curve. Such estimates are technically more challenging and probably less accurate for copolymers, because multiple ratios of monomers can yield very similar force/extension profiles. Although this ratio can be constrained somewhat by the monomer ratio determined spectroscopically for a bulk sample, the single-molecule nature of SMF experiments means that the composition of the measured macrochain will almost certainly deviate from that measured in a bulk sample. The outstanding question is the probability of such deviation as a function of its magnitude. An illustrative but inconclusive example is provided by force/extension curves of copolymers of isomeric cinnamate dimers,[25] where the composition of each stretched chain was estimated both from the number of mechanical instabilities, and the micromechanics of the chain prior to the reaction. Although the two methods appear to yield the ratios in a reasonably good agreement, the chain-to-chain variation of the composition was considerable, illustrating the caution warranted when using quantities measured on bulk samples to characterize individual chains.

The technical difficulty of modeling force/extension curves from quantum chemical calculations means that molecular interpretation of most SMF experiments relevant to polymer mechanochemistry relies on mostly qualitative arguments, and which therefore might be best viewed as tentative.

Quantitative interpretations of SMF experiments typically aim at estimating how "sensitive" the reaction rate is to force (usually by fitting the experimental observations to the Bell–Evans equation, despite the long-articulated concerns that such fits do not yield a unique set of parameters[71]) or, more infrequently, estimating the strain-free reaction rates (activation energies) of reactions that are too slow to be measurable (e.g., dissociation of various covalent bonds). Such fitted or extrapolated values were used to validate the nature of the reaction responsible for the observed chain micromechanics, or even to speculate about the structure of the transition state or the reaction mechanism. Quantitative interpretation of SMF experiments is even more challenging than qualitative interpretation and is plagued by two largely unresolved and rarely acknowledged problems: the lack of reliable means of extracting ensemble-average parameters (e.g., activation free energies, geometrical changes) from intrinsically stochastic limited statistics measurements on single molecules, and the difficulty of controlling the pulling geometry at the atomic level.

Every known model of mechanochemical kinetics relies on activation energy, which is an ensemble quantity, i.e., it is only meaningful for a sufficiently large collection of particles (e.g., reacting molecules) to average out thermal fluctuations of observed quantities (such as rate constants or survival probabilities). In contrast, the behavior of polymer chains in single-molecule experiments is governed by single-molecule statistics. To appreciate the difference, consider a hypothetical macromolecule containing 10^{12} equivalent nonscissile reactive sites (a molecule with so many reactive sites is not synthetically accessible). We'll stretch this molecule very rapidly to restoring force F and monitor by whatever means available (e.g., increase of the chain contour length) the change in its composition due to the mechanochemical reaction as we maintain the chain at this restoring force, i.e., perform a force-clamp experiment. This data can be converted to the reaction rate constant at force F, $k(F)$. If we repeat the experiment with an identical macromolecule, we may expect the two rate constants to be within $\sim 15\%$ of each other if we are competent experimentalists, the variance being a reflection of experimental error.

In practice, the vast majority of SMF experiments reported to date were performed on macromolecules containing fewer than 100 equivalent reactive sites for synthetic polymers and as few as eight sites for proteins, whose monodispersity (i.e., all chains have exactly the same length and the number of the reactive sites) eliminates one important source of experimental variability. Repeating the above experiment with a macromolecule containing only eight reactive sites will have only $\sim 7\%$ chance of exactly half of these sites having reacted within the same time $(\ln(2)/k(F))$ that exactly half of the sites in the hypothetical large analog do. Repeating this experiment 20 times increases the chance that the inferred average rate constant is between half and twice the ensemble-average value (i.e., in the $(0.5-2)k(F)$ range) to only

<50% even if all other sources of variability are eliminated, which is impossible. The fairly slow convergence of rate constants (and parameters derived from them, such as the "length" of the transition state) to the ensemble-average values means that quantitative interpretation of SFM results based on reactions of fewer than $\sim 10^3$ sites is probably unreliable until more powerful mathematical methods of extrapolating limited statistics to the thermodynamic limit have been devised.

Because many reaction rates increase exponentially with force, the rate constants derived from dynamic measurements may, in theory, converge to the ensemble limit faster[25] with the number of reactive sites than for the force-clamp experiment. In practice, this advantage of dynamic SMFS is diminished, and may even be eliminated by the dependence of the critical force in such experiments on the polymer contour length and its compliance. Dynamic force spectroscopy controls chain extension (tip retraction) rate, whereas reaction kinetics is controlled by the restoring force, and longer macromolecules thus require more time to reach the same restoring force as shorter chains. Consequently, the longer chains experience smaller effective loading rate and hence the survival probabilities of individual reactive sites decrease slower with time than in shorter chains (Eq. 7). Longer chains also have more equivalent reactive sites, and the probability of one site reacting increases as a power of the number of equivalent sites. Because of the high dispersity (index >1.5) of synthetic polymers used to date in the reported SMF experiments and limited control over the interactions between the chain and the surfaces (see next paragraph), the length of chain segments that are stretched in such experiments varied by up to fivefold among repeats. The net result is that the force at which the chain micromechanics is detectably affected by mechanochemical reactivity in repeat SMF experiments varies by hundreds of picoNewtons. These effects are not considered by any model of mechanochemical kinetics, but are amenable to numerical simulations using data obtained by quantum chemical calculations (Fig. 4).[25]

Technical idiosyncrasies of SMFS may also introduce systematic errors that thus far appear impossible to quantify, much less to eliminate. Although SMF experiments are often depicted with the stretched chain aligned with the direction of motion of the positional scanner and hence with the axis of the cantilever deflection (z direction), this alignment is highly improbable. Far more likely are geometries in which the backbone of the stretched chain forms an angle to the direction of the motion. Likewise, it's highly unlikely that the chain is connected at the apex of the cantilever tip, instead of

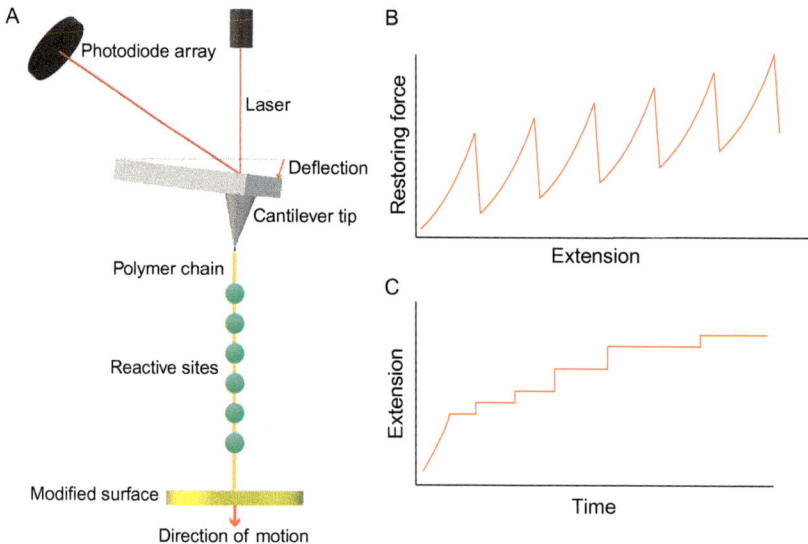

Fig. 4 Simplified illustration of an AFM-based single-molecule force experiment. A polymer chain is attached at its termini to a cantilever tip, and a reactive surface, and the latter surface is retracted. In dynamic force spectroscopy, the retraction typically occurs at a constant velocity, gradually stretching the polymer and deflecting the cantilever, which is measured by the change in the position of the reflected laser beam at the detector. In the force-clamp mode, quick retraction to achieve a desired cantilever deflection is followed by restricted movement necessary to maintain this deflection. The data from the two variants are reported either as force/extension (B) or extension/time (C) curves. When stretching destabilizes one or more monomer of the macromolecular bridge enough for it to react a mechanical instability results, producing either a sawtooth pattern (B) or an abrupt elongation of the polymer contour length (C). Thermal fluctuations of the tip and the macromolecule result in small, random fluctuations of the extension not shown in (B) and (C). The drawing in (A) is not to scale, with the AFM tip being an order of magnitude larger than the macromolecule.

somewhere on its side. Both factors (illustrated in Fig. 5) result in the chains in SMF experiments very likely being stretched to larger restoring forces than that derived from the deflection of the cantilever, because cantilevers are much easier to bend vertically than laterally and the lateral component of the restoring force is thus not measured (and may currently be unmeasurable). Since it appears impossible with the current configuration of SMF experiments for a chain to bind in a way that the cantilever deflection force exceeds the chain restoring force, repeat measurements cannot compensate for off-axis stretching, and the reported forces are probably systematically underestimated by a factor that can neither be estimated nor eliminated in repeat measurements.

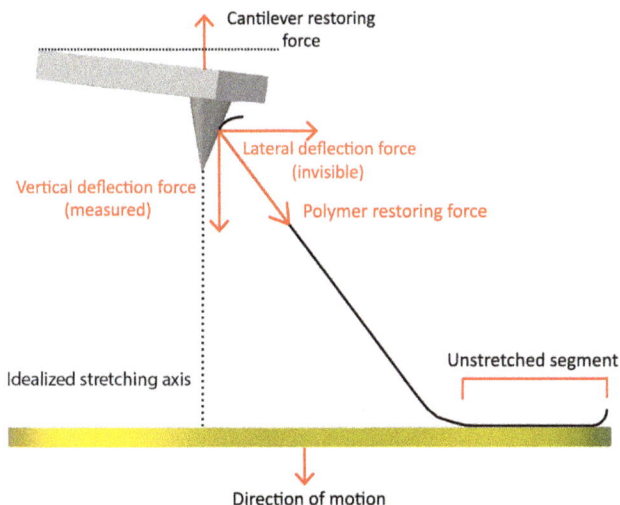

Fig. 5 Illustration of a major mechanism whereby SMF experiments underestimate the restoring force of the stretched macromolecule: the chain being misaligned with the direction of motion of the surfaces (illustrated by the *broken line*) and the chain binding to the tip off-apex. The restoring force of such a chain can be separated into vertical and lateral components, with only the former being measured due to the high stiffness of the cantilevers in the horizontal plane.

Reactions that are reversible on the timescale of an SMF experiments allow the same transition to be observed multiple times within the same macromolecule by performing repeated cycles of retraction/return. Such cycling would allow statistical averaging of the measured kinetics without variability of stretching multiple macrochains or the need for multiple equivalent reactive sites. It would also allow experimental testing of the Jarzynski equality[72] on reactants whose ensemble-average reactivity can be readily quantified or computed quantum chemically. Unfortunately, so far, no reaction of interest to polymer mechanochemists has been identified that is reversible on the timescale of an SMF experiment. Isomerization of spiropyran to merocyanine should be reversible based on the estimated strain-free kinetics[73] but stretching of macromolecules containing two different spiropyran derivatives yielded irreversible isomerization for unknown reasons.[74]

Optical and magnetic tweezers, in which a macromolecule is bound to one or two microbeads whose position(s) are manipulated by intense focused electromagnetic fields have been utilized to stretch biopolymers and quantify the forces generated by various motor proteins.[75] These techniques generate forces < 100 pN (vs up to ~ 5 nN by AFM), which is insufficient to accelerate reactions involving covalent bond rearrangement to the second timescale. Consequently, optical and magnetic tweezers have not been used in polymer mechanochemistry.

3.2 Flow Fields

A number of experimental techniques exist that rely on hydrodynamic coupling between a polymer solute and solvent in a flow to stretch the macromolecules. It is practically impossible to generate a fluid flow with flow rate that is uniform in space. A velocity gradient perpendicular to the flow direction (which is realized in a fluid flowing past a surface, whereby the solvent flow increases from zero immediately at the interface (stagnation layer) to a maximum far away from the surface) creates shear. A velocity gradient along the direction of flow produces elongation. The strength of the flow is quantified by a strain rate, $\dot{e}$, or shear rate, $\dot{\gamma}$, which reflects flow velocity gradient along or perpendicular to the flow direction, respectively, per unit length. In practical flows both components are present, although certain portions of flows may approximate pure shear or pure elongational flows. Quasi-steady-state elongational flows are produced by the filament-stretching device[76] (uniaxial flow), cross-slot, and four-roll mill devices[77] (planar flows). Transient elongational flows are generated in abrupt-contraction devices and by ultrasonication[9,15] (see Section 3.2.2). In all these geometries turbulent flows are also present, and their contribution to the observed mechanochemistry is rarely known.[78] Although the rapid and chaotic changes in flow velocity of turbulent flows largely preclude elucidation of the microscopic conditions responsible for mechanochemistry, they are studied in large part due to the industrial importance of using macromolecules for drag reduction, as polymer degradation severely reduces the efficiency of this process with time.[1]

Two key parameters determine the fate of a macromolecular chain in a dilute solution under flow: the Deborah or Weissenberg number, and accumulated (or Hencky) strain.[79] The former is a product of the strain or shear rate and the longest relaxation time of the polymer, τ_1. In flows with rates below approximately half $1/\tau_1$ the polymer solute remains in its coiled geometry, only marginally affected by the flow. The longest relaxation time increases linearly with the solvent viscosity and as a power of 1.5 of the polymer contour length. Synthetically accessible polymers in common organic solvents have relaxation times in the 100 ns to 10 μs range (e.g., τ_1 of 1 MDa and 100 kDa polystyrene in THF at 300 K are 7 and 0.2 μs, respectively).

Quasi-steady-state elongational flows with strain rates on the order of $10^4\,s^{-1}$ have never been demonstrated experimentally, making such flow geometries of limited value for studying mechanochemistry of synthetic polymers (synthetic polymers can be stretched when dissolved in special high-viscosity solvents, e.g., Boger fluids, which introduce their own complications[79] and have not been used in polymer mechanochemistry). In contrast, DNA polymers with relaxation times in the seconds range are readily available and such chains can be stretched easily even in disposable PDMS-made cross-slot devices.[80]

In theory, a macromolecule in elongational flow with a strain rate exceeding $0.5/\tau_1$ would undergo abrupt coil–stretch transition, whereby the end-to-end distance of the chain increases close to its contour length (in elongational flows) or to a substantial fraction of it (in shear flows). In practice, the coil–stretch transition is quite slow and requires a substantial residence time in the flow, which is quantified by the Hencky strain. For a flow with time-independent strain or shear rate, Hencky strain is a product of this rate and the residence time. For Hencky strain below a certain threshold value, which depends on the Deborah number, stretching is highly transient, and the observed macromolecular conformations vary greatly from one molecule to another.[79] It seems safe to speculate that polymers of interest in contemporary polymer mechanochemistry cannot be fully stretched in any experimentally realizable flows, and any discussion of the molecular origin of the observed reactivity of such solutes should acknowledge that bulk response arises from a broad (and currently unknown) distribution of conformers (Fig. 6).

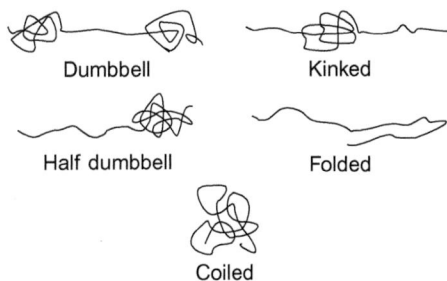

Dumbbell   Kinked

Half dumbbell   Folded

Coiled

Fig. 6 Cartoon representation of the types of DNA conformers thought to occur in quasi-steady-state-elongational flows at Hencky strain corresponding to undetectable amount of coiled chains. *Based on data from Perkins TT, Smith DE, Chu S. Single polymer dynamics in an elongational flow. Science. 1997;276(5321):2016–2021; Hur JS, Shaqfeh ESG, Larson RG. Brownian dynamics simulations of single DNA molecules in shear flow. J Rheol. 2000;44(4):713–742.*

3.2.1 Quasi-Steady-State Elongational Flows

Quasi-steady-state elongational flows are characterized by the existence of a central stagnation point with zero net fluid velocity. Such flows are typically induced in either a cross-slot device (Fig. 7)[81] or four-roll mill.[77] The solvent strain rate is controlled either by the pressure difference between inlet and outlet channels in the cross-slot device, or by the speed of the rollers. A dissolved macromolecule is occasionally trapped at the stagnation point, where (at least in theory) it can be kept for hours with an adequate feedback mechanism. In practice, the achievable residence time depends on the desired strain rate and decreases very rapidly at the strain rates relevant in polymer mechanochemistry because of the difficulty in maintaining the required flow stability rather than mechanochemical fragmentation.

Indeed, no convincing evidence of mechanochemistry in polymers stretched in quasi-steady-state elongational flow has ever been reported. Simulations suggest that early observations of polymer fragmentation in such flows appear to result from chain fragmentation in turbulent flows at the edges of the cross-slot device.[78] The primary contribution of QSS

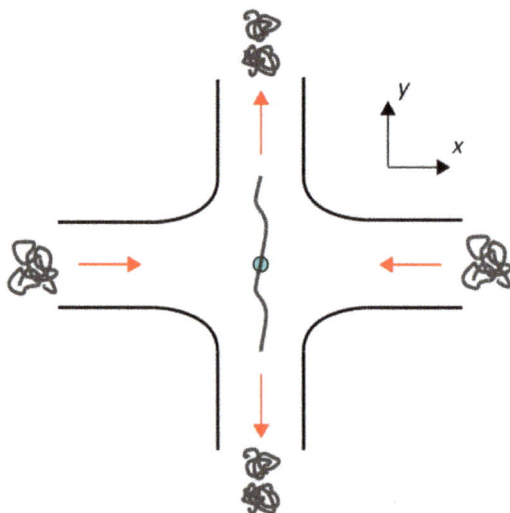

Fig. 7 Illustration of the cross-slot device used to study polymer extension and degradation in QSSF. Two opposing channels along the $\pm x$ directions pump solvent containing a single macromolecule into the device, and two opposing channels along the $\pm y$ directions suck out the solution. The strain rate experienced by the trapped macromolecule is controlled by changing the velocity of the pumped solution. The central stagnation point is marked with a *pale blue circle*. In this region, net fluid velocity along either axis is zero.

elongational flow to the current discussions of polymer mechanochemistry is a force distribution along a chain trapped at the stagnation point, which is often erroneously assumed to represent force distribution along a chain stretched in any flow field.[1] This force distribution is derived from the simplest possible implementation of the classical bead–spring model,[79] which represents a polymer chain as a series of spherical beads connected by harmonic springs. This implementation neglects hydrodynamic interactions, which account for the distortion of the flow field by the beads; excluded volume interactions, which account for repulsive or attractive bead–bead interactions not mediated by the springs, and thermal fluctuations, which ensure that a chain geometry is more complex than that of a rigid rod. These assumptions result in a quadratic dependence of the force experienced by a bead (or equivalently the restoring force of a spring attached to it) on its position relative to the center of mass of the chain, independent of the parameters of the model (radius of the beads, friction coefficient, and spring force constants). The same dependence is obtained if the chain is assumed to be a rigid slender rod in a 1D solvent flow.

This "rigid-rod" model is almost certainly irrelevant to any experimentally observed mechanochemistry in flows, simply because the underlying assumptions are too unrealistic. While the importance of hydrodynamic or excluded volume interactions may not be obvious, the assumption of the chain being in internal mechanical equilibrium (i.e., behaving as a slender rigid rod) seemingly requires Hencky strains that are unachievable with any experimentally demonstrated flows and polymer contour length relevant for contemporary studies in polymer mechanochemistry. The application of this rigid-rod model for sonication is sometimes justified by arguing that a parabolic force distribution along the chain is required to explain the "mid-chain" scission of polymer chains in flows. The latter refers to a common observation that the molar mass distribution (MMD) of the product of chain fragmentation in flows has the maximum at the chain mass approximately half that of the original polymer. This "mid-chain" scission, however, is consistent with an arbitrarily large number of force distributions along the chain,[1] and the argument unfortunately conflates a macroscopic observation (MMD averaged over fragmentation of many molecules comprising the bulk sample) with the microscopic conditions (the distribution of fragmentation probability along an average fragmenting chain) responsible for it. Imagine that contrary to the rigid-rod model described earlier, only a small segment of each fragmenting chain is stretched (i.e., fragmenting chains resemble those in Fig. 6 instead of rigid rods) but that any portion of the chain has a nonzero probability to be stretched. The symmetry argument alone suggested that this probability will be higher for segments closer to the chain center of mass than those farther away. When averaged over an ensemble of chains, the most probable product of fragmentation will be half of the original mass even if every backbone bond of the stretched segment of individual fragmenting chains has the same restoring force and the same fragmentation probability.

3.2.2 Sonication

Sonicating a dilute solution of a polymer is by far the most popular technique of polymer mechanochemistry. It is as easy to perform technically as it is difficult to understand at the microscopic or molecular levels, or indeed quantify reliably.

In sonication, transient elongational flows needed to stretch a macromolecular solute are generated by the collapse of cavitation bubbles. Passing sound waves of frequencies in the kiloHertz range through a liquid creates acoustic cavitation, which is the nucleation, growth, and subsequent collapse of cavities within solution (bubbles). The acoustic field of the ultrasound waves dilates and contracts the bubbles, eventually causing them to collapse violently. Such a collapse of an isolated cavitation bubble creates a spherically symmetrical transient elongational flow field with fluid elements closer to the bubble edge having higher velocity than those farther away, as illustrated in Fig. 8.[15]

Sonication of commercial polymers, such as polystyrene or polyacrylates, results in their fragmentation as evidenced by a gradual decrease in the average molar mass of the solute as sonication progresses. Chain fragmentation first produces highly reactive macroradicals, which are probably quenched by a reaction with the solvent or (more likely) sonolytically generated small-molecule radicals, i.e., species produced inside the collapsing bubbles from solvent vapor or dissolved gases, but not the polymer solute, which cannot enter the bubbles. Recombination of macroradicals is likely to be negligible because no evidence of a reaction between a macroradical and a polymer chain has ever been reported (e.g., sonication of a polymer containing sp^2 C atoms could be expected to create a product with higher molar mass than the original reactant due to high reactivity of macroradicals toward addition to sp^2 C atoms). The very limited chemistry possible upon stretching simple polymers such as polystyrene makes sonication of their solutions of limited interest in polymer mechanochemistry (a topic of some interest is the scaling of the bulk rate of mechanochemical fragmentation of a

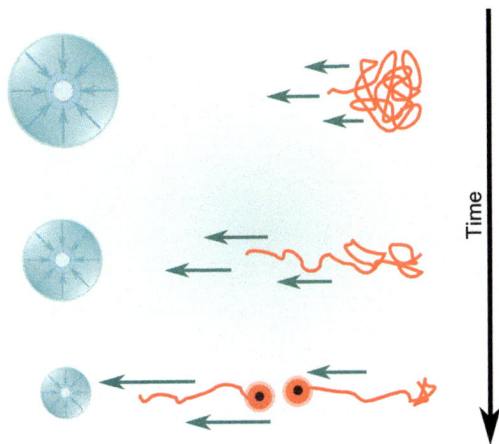

Fig. 8 The mechanism by which ultrasonic degradation of a polymer chain is thought to occur during sonication. At the moment of bubble collapse (*top*), the polymer chain is coiled in solution and its instantaneous shape is roughly spherical. As the cavity rapidly decreases in size, the surrounding solvent molecules are drawn toward it, and the segments of the polymer chain closest to the bubble are stretched. At the latter stages of bubble collapse (*bottom*), the solvent strain rate is sufficient for the chain to fragment. *Arrows* show the direction of solvent motion at each point, with the magnitude of its velocity represented by its size.

polymer on its molar mass, which is reviewed in Ref.[1]) Instead, contemporary focus has been on sonicating solutions of polymers comprised of one or multiple reactive sites embedded in inert polymer backbones. In most cases these reactive sites are dissociatively more labile than the rest of the polymer backbone bonds, in theory resulting in the stretched polymer fragmenting preferentially at the reactive site (site-selective fragmentation) instead of elsewhere along the backbone (nonselective fragmentation). In practice, both selective and nonselective fragmentations are detectable during sonication. Examples of scissile reactive sites are Diels–Alder adducts and cinnamate dimers. Two types of nonscissile reactive sites studied to date are dihalocyclopropanes and spiropyrans.

Sonication of solutions of such polymers yields some of the same products that are thought to be generated in single-molecule stretching of isolated macrochains or in bulk polymer samples under macroscopic loads. Sonication of polymers containing multiple equivalent reactive sites allows accurate spectroscopic identifications of the products of sonication,[32] something that is not possible in SMF experiments and is rarely done for solid-loaded polymer samples. UV–vis and/or fluorescence spectroscopy is particularly valuable in confirming site-selective fragmentation if the studied reaction is mechanochromic, i.e., it yields a product with optical properties distinct from the reactant. In contrast, a reduction in the average molar mass of the sonicated sample can be caused by both selective and nonselective fragmentation and hence cannot be used to distinguish between these two paths (for a recently demonstrated example, see Ref.[6]).

Consequently, sonicating a polymer solution is an easy and qualitatively reliable way of testing if the kinetic stability of a particular reactive site is affected by stretching it, as long as the products of sonication are amenable to spectroscopic characterization. In other words, sonication is useful to confirm force acceleration of both scissile and nonscissile mechanochromic reactions (i.e., regardless of whether they result in chain fragmentation or not), and of nonscissile reactions that occur at multiple equivalent reactive sites per chain. In contrast, acceleration of nonmechanochromic scissile reactions to an extent greater than that of the "inert" backbone bonds cannot generally be established reliably by sonication. Sonication of chains containing multiple scissile reactive sites yields fragmented chains with most reactive sites intact, because once the chain fragments by the dissociation of a single site, the probability of it getting stretched enough to accelerate the dissociation of another site on the experimental timescale becomes negligible. As a result, NMR spectra of such sonicated polymers are dominated by the intact reactive sites, precluding the quantitation of the reaction extent. Several workarounds are known, but none allows the extent of the reaction to be estimated, i.e., it is impossible to establish if the mechanosensitive reaction is accelerated by force to any greater degree than dissociation of the "inert" bonds of the backbone, which always occurs during sonication.

Every reactive site that has ever been incorporated in an "inert" polymer backbone in context of polymer mechanochemistry is, in the absence of force, more labile (often by many orders of magnitude) than the dissociation of the backbone bonds. Of particular interest are thus examples where the macromolecules manifest no preference for reaction at such reactive site over fragmentation at the adjacent backbone bonds, because such cases are rare and challenge the "intuition" of how the kinetics of molecular fragmentation responds to tensile load. Several reactions appear to be insensitive to tensile load when the respective polymers are sonicated but are accelerated, albeit to a small degree, in bulk loaded materials. The difference probably reflects the highly transient nature of chain stretching in sonication, which dictates that only reactions over barriers <10 kcal/mol ($t_{1/2} \sim 1\ \mu s$) are observed. The longer relaxation times of chains in solids probably mean that a stretched polymer segment may remain in its nonequilibrium geometry for longer than a few microseconds, allowing reactions over barriers in excess of 10 kcal/mol to be observed.

Sonication has been shown to initiate reaction cascades, whereby a primary stable product of mechanochemical reaction reacts with another component of the sonicated solution. Such experiments are valuable in supporting the existence of similar cascades in bulk materials under load, where they have to be exploited to yield mechanochromic and load strengthening responses.

Practically nothing is known about the microscopic conditions responsible for polymer mechanochemistry in sonicated solutions. Our current understanding of the dynamics of isolated cavitation bubbles is quite sophisticated, but no attempt to establish the feasibility of mechanochemical reactions in a flow field generated by an isolated collapsing bubble has ever been reported. Instead, polymer solutions are sonicated using macroscopic immersion acoustic horns that produce an indeterminate number of cavitation bubbles, hundreds or thousands of which may entrain to form bubble clouds with poorly understood dynamics. Under these conditions, the observed chemistry may not even result from the collapse of individual bubbles, but rather from shock waves created by a synchronized collapse of bubble clouds. A simple hand-waving argument suggests that the conditions experienced by mechanochemically reacting chains in sonicated solutions are quite extreme. The low limit of the maximum fluid strain rate that is generated during sonication is based on the observation that even chains as short as $\sim$20 kDa polystyrene fragment during sonication, which requires that they reside, at least transiently, in flow fields with straining rates of $>0.5/\tau_1 \sim 3 \times 10^6 \, s^{-1}$. At the most utilized sonication frequency of 20 kHz, a bubble must collapse within <25 μs (the duration of a single compression cycle) with the fluid strain rates $>10^6 \, s^{-1}$ generated only at the final stages of collapse. This means that a chain segment can go from a strain-free geometry to being stretched to several nanoNewtons of force in <1 μs, a loading rate that is $\sim 10^5$ times greater than can be generated in SMF experiments. A polystyrene chain has $\sim$50% survival probability against fragmentation within 1 μs at 5 nN, which gives a very approximate upper limit of the extent of strain a chain experience during sonication. An average chain may be stretched multiple times before fragmenting.

Few attempts to estimate the microscopic conditions responsible for mechanochemistry in sonicated solutions have been reported. To date, the most useful insights have come from analysis of the fraction of mechanochemically reacted chains that have undergone site-selective chemistry vs nonselective fragmentation. While this effort remains in very early stages, quantifying the microscopic conditions in sonicated solution is likely the most important factor in realizing the full potential of sonication to understand chemical reactivity at extreme strains and loading rates.

Although most reports of sonication of polymer solutions contain some characterization of the rate at which the bulk composition of the sonicated solution changes with sonication time, both technical and fundamental limitations ensure that such data are far less valuable in gaining the molecular insights than kinetic studies in physical organic chemistry generally are.

The fundamental reason is that macroscopic rate constants of a sonicated solution are a complex and unknown convolution of multiple microscopic probabilities. In conventional kinetic studies, a rate constant reflects the microscopic probability of a molecule to react because every reacting molecule scales the same energy barrier and the probability of doing so is determined by the Boltzmann distribution of the reactant molecules. In contrast, at any moment in a sonicated solution, different macromolecules have to traverse vastly different activation barriers to undergo the same reaction, based on how much each is stretched by the flow field, how fast this field changes in time, the history of the molecule in this field, and the probability that the molecule will remain in the field among other contributions.

The practical difficulties are related to the finite dispersity of polymer samples, i.e., they are comprised of chains of (sometimes vastly) different lengths, and hence propensity to undergo mechanochemistry in flows. A consequence of this dispersity is the need to discuss the composition of a polymer sample in terms of its MMD, rather than a small number of components. Sonication of a polymer solution fragments polymer chains, changing the MMD of the solute. In theory, these time-lapsed MMDs can be used to estimate the distribution of fragmentation probabilities along the polymer chain, which can allow various assumptions about the microscopic conditions responsible for the observed bulk changes to be assessed. In practice, such analyses are technically complex. As a result, most reports in the literature reduce each MMD to a single distribution moment (e.g., number-average or mass-average molar mass) and then fit a set of these moments to one of a plethora of empirical rate laws proposed in the literature to describe the time evolution of the molar mass of a sonicated polymer. This procedure discards important details about reaction kinetics. Worse still, the resulting fitting parameters from most such models (including the most commonly used one now) are neither mechanistically significant nor allow comparisons across different experiments, because the underlying models were derived either from an internally inconsistent set of assumptions or for conditions too dissimilar to those for which they are applied. An example of the former is to simultaneously postulate that the probability of a backbone to break decreases linearly with the number of backbone bonds, and that chains with fewer than a minimum number of backbone bonds do not break. An example of the latter is to fit the rate of molar mass reduction of the polymer claimed to break only at a single site to a model that postulate that each backbone bond has the same fragmentation probability.[1]

Any reaction induced by stretching a polymer chain for a 1 μs or less is likely to follow pseudo-first-order kinetics. This in theory would eliminate the need for empirical models of MMD evolution and would allow the kinetics to be described by the ensemble-averaged first-order rate constant of the depletion of the total mass fraction of the chains comprising the original reactant. In practice, most sonication experiments have been conducted with polymers of such large dispersity that the MMDs of the reactant and the product cannot be quantified individually, as they overlap. Many reported sonication experiments were conducted on samples in which the average molar mass of the largest 10% of the chains was 10-fold or greater than the average molar mass of the smallest 10% of the chains. MMDs measured on such samples manifest clear evidence of preferential depletion of the higher-mass component, which produces fragments of mass indistinguishable from that of a fraction of the reactant, further complicating molecular interpretation of the kinetics of changes in the bulk composition of the sonicated sample.

All these problems can be overcome by using very low-dispersity polymers (PDI <1.005), high-resolution size-exclusion chromatography, and cleverly designed reactive sites that allow one to differentiate whether multiple products of sonication results from intra- vs interchain kinetic competition. Such experiments, however, remain to be reported.

3.3 Mechanochemical Reactivity Without Macroscopic Motion

Mechanochemical reactions are manifestations of molecular strain induced by macroscopic motion, such as the macroscopic flow of a fluid, translation of an AFM tip, or distortions of the macroscopic dimensions of a bulk polymer sample. Atomistic studies of such processes are particularly challenging. Two broad approaches suggest that the effect of anisotropic molecular strain on chemical reactivity can be studied productively without the complexities of macroscopic motion: one relies on small-molecule macrocycles and the other on overcrowded polymers that spontaneously stretch at certain interfaces. The latter was reviewed in detail recently[82]; the other is briefly analyzed below. Because molecular strain (unlike its engineering counterpart) is a qualitative concept, no rigorous definition of what makes molecular distortion anisotropic is possible. We find a functional definition to be informative: a molecular fragment is axially strained if its geometry deviates from that of the same molecular fragment in a stretched macromolecular segment by less than a threshold RMS value. Such comparisons require the two structures to be optimized quantum chemically and are thus only as reliable as the chosen model chemistry.

One approach reported by our group in 2009[53,55] is to dispense with macromolecules altogether, and attempt to reproduce in a strained macrocycle the distortion that a reactive site in the backbone of a stretched macromolecule experiences (Fig. 9). Evidence accumulated since then clearly demonstrates

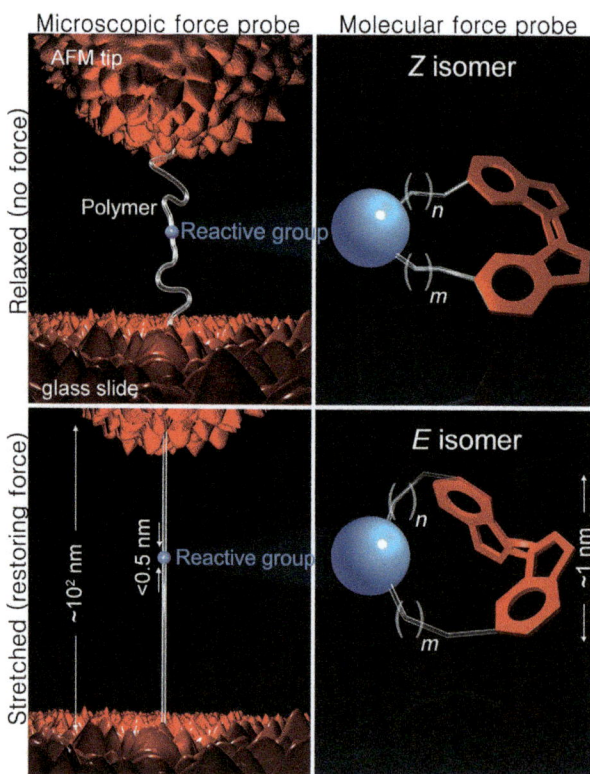

Fig. 9 Schematic comparison of an SMF experiment on a polymer chain (*left panels*) containing a reactive site (*blue*) with an experimental model using stiff stilbene (*right panels*). *Reproduced with permission from Yang QZ, Huang Z, Kucharski TJ, Khvostichenko D, Chen J, Boulatov R. A molecular force probe. Nat Nanotechnol. 2009;4(5):302–306.*

that the approach does not simply succeed in reproducing in a small molecule the reactivity observed upon stretching polymers containing the same reactive sites,[58] but also offers insights into mechanochemical reactivity not attainable by any other experimental technique.[42,61]

The approach uses E stiff stilbene to impose an approximately axial tensile strain on a reactive site connected to the C6, C6' carbon atoms of stiff stilbene by short inert linkers. The magnitude of the imposed strain is controlled by the length and the conformational flexibility of these linkers, producing a series of increasingly strained macrocycles in which the reactive site geometry approximates that in macromolecules stretched to between 100 and 800 pN, in increments as small as 50 pN. The Z isomers of these macrocycles are strain free regardless of the linkers and are valuable both as efficient synthetic precursors of the strained isomers, and as strain-free references. Comparing the reaction kinetics in the two isomers of the same macrocycle allows the effect of the anisotropic strain to be isolated from any other influences, including the linkers and the solvent. In other words, stiff stilbene acts as a molecular force probe, in analogy to the microscopic force probes of SMFS (Fig. 9).

The ability to study quantitatively the effect of anisotropic strain on localized chemical reactivity in small molecules instead of stretched macromolecules offers practical and conceptual advantages. First, small-molecule reactivity, including reaction kinetics, selectivity, and mechanisms, is amenable to detailed characterization by the full complement of experimental and theoretical tools of modern chemistry. Second, unlike SMF experiments, molecular force probes are studied in ensembles, obviating the need to estimate ensemble properties from single-molecule statistics. Third, unlike polymer mechanochemistry in flow fields, all reacting molecules are subject to the same uniform conditions, and macroscopic reaction rates reflect the microscopic reaction probabilities. Fourth, molecular force probes are suitable to quantify mechanochemical kinetics that is beyond the scope of other techniques, including reactions that are strongly inhibited by force or even just too weakly accelerated by force to compete with chain detachment in SMF or chain fragmentation in flow fields. Finally, because the entirety of the macrocycles is amenable to full atomistic description at a quantum mechanical level, molecular force probes allow the key assumption of mechanochemical kinetics, that the effect of many molecular degrees of freedom on localized kinetics is captured quantitatively by a single parameter with the meaning of molecular restoring force, to be evaluated with unprecedented detail.

For the observed reactivity trends in series of stiff stilbene macrocycles to advance our understanding of polymer mechanochemistry, they must be expressed as a function of the restoring force of a local coordinate of the reactive site (which then can be related to the single-chain force as described in Section 2.2). The physical meaning of this force is the same in the macrocycles and in stretched polymer chains: a quantifier of kinetically significant molecular strain energy of a portion of the molecule. At stationary states, whose relative energies determine the reactivity, the molecules are in internal mechanical equilibrium with zero force on every atom. In other words, no Hamiltonian exists whose eigenvalue is a restoring force to a molecular coordinate of a macrocycle, and estimates must rely on one of several existing models.[1] A similar reliance on a model underlies the single-chain stretching forces reported in SMF experiments, where the restoring force of the stretched macromolecule is neither controlled nor measured directly, but rather estimated from the deflection of the cantilever. Contrary to common belief, these estimates are subject to both systematic and random errors, including the lack of molecular-level details or control over the chain/surface interaction (see Section 3.1) and the limits of the simple models used to relate the measured deflection to the force (e.g., the beam equation) to capture such dependence accurately, particularly at high forces. In other techniques of polymer mechanochemistry, including flow fields and bulk materials under loads, the magnitude of the distortion of the chain responsible for the observed reactivity cannot be estimated at all.

The two main contributions of molecular force probes to our understanding of polymer mechanochemistry are the validation of the local assumption of mechanochemical kinetics (see Section 2.2) and experimental demonstration of the diverse range of responses of molecular fragmentation kinetics to axial tensile strain. In addition to the conventional notion that stretching a molecule accelerates its fragmentation along the stretching axis (which follows the Bema–Hapothle, or free energy relationship (FER) postulate) molecular force probes provided examples of reactive sites that are kinetically stabilized against fragmentation along the stretching axis or that are kinetically destabilized toward fragmentation orthogonal to the pulling axis.[42] Neither pattern is consistent with an FER (either linear or quadratic), because each corresponds to an aphysical value of the normalized reaction coordinate for the position of the rate-determining transition state, α. A perturbation (e.g., stretching) that both inhibits a reaction and lowers its standard free energy requires $\alpha < 0$ and one that accelerates a reaction without affecting its standard free energy (or enthalpy) requires $\alpha = \infty$ (by definition, $\alpha = 0$ for the reactant and $\alpha = 1$ for the transition state). A quadratic FER (e.g., the Marcus equation) accommodates $\alpha < 0$ but postulates an "inverted region" if the reaction energy exceeds a threshold value and still requires $0 < \alpha < 1$ for all other values of the reaction energy. In contrast, a nonscissile mechanochemical reaction that is inhibited by stretching force has $\alpha < 0$ irrespectively of the reaction energy and even a self-exchange reaction has $\alpha < 0$. Furthermore, in these mechanochemical reactions the reaction energy is independent of the force, whereas the activation energy is.

In contrast, the local coordinate approximation of mechanochemical kinetics (see Section 2.2) both adequately rationalizes all demonstrated types of responses of the kinetics of molecular fragmentation to tensile strain and supports the design of new reactive sites that are likely to follow each of these patterns.

In addition to serving as molecular force probes, the capacity of stiff stilbene to impose anisotropic molecular strain has been successfully exploited in self-assembly,[83] molecular motors,[84] and catalysis control.[85] Proposed or speculated applications of stiff stilbene include thermal storage of solar energy[2] and molecular photoactuation.[86] Unfortunately, a recent claim that at least highly strained E macrocycles may cause nonstatistical reaction dynamics of the attached reactive sites is highly unlikely to be correct.[87] In the vast majority of chemical reactions, the reactant(s) remain in thermal equilibrium with its environment, and to a very good approximation, the reaction probability is proportional to the fraction of molecules with energies in excess of the rate-determining activation barrier.[13] In nonstatistical reaction dynamics,[88] which occurs in some gas phase reactions and potentially in conformational rearrangements of some proteins in solution, the reactant is not in thermal equilibrium with its environment and the vibrational temperature of an average reactant molecule exceeds the temperature of its thermal bath. This excess energy generally results from a photon absorption (as stiff stilbene does at ~375 nm) or a preceding exergonic reaction. In solution such molecules dissipate this energy by vibrational energy relaxation (VER) at the subpicoseconds timescale.[87] In theory, a highly vibrationally excited (hot) molecule can traverse a sufficiently small activation barrier before it thermalizes, in which case the observed rate will far exceed the one predicted by the TST. In practice, no molecule larger than ~10 atoms in solution has ever been shown to manifest nonstatistical reaction dynamics.

The two main disadvantages of the molecular force probes relative to their microscopic analogs in SMFS for studying mechanochemical reactivity are the meaningfully smaller maximum force that they can impose on the reactive site, and the maximum size of the reactive site that can be stretched by stiff stilbene. The former is determined by the kinetics of thermal $E \rightarrow Z$ relaxation and the synthetic difficulty of obtaining highly strained E macrocycles. The same mechanism that lowers the kinetic stability of many (but not all) reactive sites incorporated in E macrocycles also lowers the kinetic stability of the E stiff stilbene itself. The activation free energy of $E \rightarrow Z$ isomerization is lowered from ~42 kcal/mol in strain-free stiff stilbene to ~16 kcal/mol at ~750 pN,[53] which makes such highly strained E macrocycles isolable only at impractically low temperatures. Likewise, such highly strained macrocycles are synthetically hard to access. The simplest means of generating E macrocycles is by irradiation of strain-free Z analogs at ~375 nm, which photoisomerizes stiff stilbene with a quantum yield that decreases almost linearly with the strain energy of the resulting E macrocycle. Because stiff stilbene is only weakly photochromic, the photostationary states of smaller macrocycles contain only a small fraction of the E isomers, which complicates their isolation and characterization.

4. BRIEF ANALYSIS OF EMPIRICAL RESEARCH IN MECHANOCHEMISTRY

Acceleration of ~20 distinct reactions in stretched polymers has been demonstrated so far, mostly by sonication. These are summarized in Ref.[1] Dissociation of ladderenes to oligoacetylene (which is related to the well-known mechanochemical [2 + 2] cycloreversion[25,63,89]) is probably the most noteworthy new mechanochemical reaction reported[10] since. At least six such reactions have been demonstrated to occur both in sonicated solutions and in mechanically loaded (usually axial compression or grinding) bulk samples, confirming that sonication mimics at least qualitatively the behavior of polymer chains in loading scenarios that are technologically more relevant but technically more challenging to study than sonication. Isomerization of dihalocyclopropanes, benzocyclobutene, and spiropyrans, and [2 + 2] cycloreversions have been studied by SMFS, in sonicated solutions and in bulk materials.

By the exacting standards of modern physical organic chemistry, our understanding of these reactions is poor. About half of all known mechano-chemical reactions were demonstrated only once, using a single mode of loading (e.g., sonication or axial loading of bulk samples) and the assumption of mechanochemical activation, while plausible, lacks credible support from quantum chemical calculations. Kinetics or even selectivities of many reactions remain to be quantified. Reaction mechanisms remain largely hypothetical even for the most extensively studied examples. The situation in large part attests to the fact that contemporary polymer mechanochemistry is very much an emerging field still in the exploratory stages with far more effort devoted to learning what is possible rather than why it is possible.

4.1 Much Ado About Dissociation of the Disulfide Bond

The kinetic stability of the disulfide bond toward either homolysis or nucleophilically assisted heterolysis (S_N2 displacement) has been a subject of a surprising number of reported studies, both experimental and computational. Homolytic S—S bond scission may be important in determining the behavior of vulcanized rubbers under load. Thiol/disulfide exchange has been studied extensively by physical organic chemists as a model S_N2 reaction, has been used widely as dynamic cross-links in polymeric materials and is important in biochemistry.

$$RSSR' + R''S^- \rightarrow RSSR'' + R'S^-.$$

Thiol/disulfide exchange (the reaction above) is an elementary (single-step) reaction that proceeds through a classical S_N2 pseudotrigonal bipyramidal transition state. Fernandez et al. reported[90] that stretching a titin containing engineered disulfide bonds in neutral aqueous solution of dithiothreitol accelerated disulfide bond reduction ~2-fold per 100 pN of applied force, a relatively small acceleration by the standards of polymer mechanochemistry. Further SMF experiments by the same group using different small-molecule reductants produced broadly similar results.[91] Interestingly, the deduced force-rate correlation extrapolated to zero force yielded a strain-free rate constant ($\sim 6.5 \text{ M}^{-1} \text{ s}^{-1}$) that is similar to those reported for the DTT reduction of disulfide bonds in several folded proteins, where the disulfide bonds reside in a fairly hydrophobic local environment and are thought to be relatively inaccessible to the solvent and hence the polar reductant solute (e.g., α-chymotrypsinogen A, $k = 9 \text{ M}^{-1} \text{ s}^{-1}$).[92] In contrast, this extrapolated strain-free rate constant is 10–200 times smaller than those for the reduction of small-molecule organic disulfides and ~10 times smaller than those in proteins with solvent-accessible disulfide bonds (e.g., trypsinogen at $>50 \text{ M}^{-1} \text{ s}^{-1}$) under comparable conditions.[93]

This trend in rate constants suggests that the acceleration of thiol/disulfide exchange observed upon stretching titin more likely reflects force-induced conformational changes in the protein environment than the intrinsic sensitivity of the disulfide moiety to tensile strain, which is independent of its surroundings. In these SMF experiments, titin was partially unfolded by subjecting it to stretching force of 130 pN for 1 s, but this "prestretching" does not eliminate the possibility that protein residues continue to dominate kinetically significant force-dependent variations in conformational compositions of the reactant and/or transition states at larger forces. Likewise, the conformational complexity of a polypeptide makes its chemomechanical coupling coefficient (i.e., the fraction of the applied force that is transmitted to the reactive site) far more sensitive to force than a simple hydrocarbon. The documented importance of force-dependent conformational changes in other S_N2 reactions involving even very short polymer fragments (e.g., neutral methanolysis of *Pr* vs Me derivatives of dialkyldiphenylsiloxane, R_2SiPh_2,[42] or basic hydrolysis of ethyl vs methyl derivatives of tetraalkylpyrophosphate, $(RO)_2P(=O)(O)(RO)_2P(=O)$[57]) at forces up to 1 nN renders the idea that a protein serves simply as an innocent transmitter of applied force fantastical.

Experimental estimates of intrinsic force/rate correlation of simple alkyl disulfide using molecular force probes and quantum chemical calculations of force-dependent activation free energies of this reaction agree that the kinetics is insensitive to force below 500 pN.[54] The result was rationalized by observing negligible elongation of the disulfide moiety in the transition state along the pulling axis. In contrast, the same methodology revealed that reduction of the disulfide moiety by phosphines in water is accelerated by force,[61] albeit weakly and by a complex mechanism, a conclusion that qualitatively agrees with SMF experiments.[91]

Several reported molecular dynamics simulations of mechanochemistry of thiol/disulfide exchange using the BLYP functional produced contradictory conclusions.[94–98] The considerable technical challenge of accurately reproducing experimental kinetics in MD simulations is illustrated by the fact that the most sophisticated to date MD simulation of an S_N2 reaction of disulfide overestimated the measured activation free energy by 1.5-fold or 10 kcal/mol (vs an error of 1 kcal/mol for static calculations[54]). Although these calculations generally aim at reproducing trends rather than absolute values, this large error suggests that the available computational methodology is not yet capable of correctly capturing the stereoelectronic factors that determine the activation barriers of S_N2 reactions at S[99,100] and the role of explicitly modeled water.[101] It is probably not justified at present to think that this error remains constant in magnitude as the moiety is distorted and hence is factored out in the predicted trend. Importantly, several of these studies reported that accelerated thiol/disulfide exchange was associated with a very unusual conformer of the disulfide moiety with a C-S-S-C torsion of 180 degree (vs the equilibrium value of ~90 degree). At higher levels of theory, the structure with the 180 degree torsion remains a transition state for rotation around the S—S bond up to at least 2 nN, suggesting that lower forces do not accelerate thiol/disulfide exchange in small organic disulfides, in agreement with the molecular-probe results.

The totality of the available data seems to suggest that at forces <0.5 nN the kinetics of thiol/disulfide exchange is far more sensitive to force when the S—S bond resides in the backbone of polymers with complex ternary and quaternary structures (e.g., polypeptides or bottlebrush polymers[82]) than in a simple organic disulfide. The origin of this difference remains to be established.

In contrast to the controversy surrounding mechanochemical kinetics of nucleophile-assisted S—S bond dissociation, force definitely accelerates homolysis of the S—S bond, including in bottlebrush polymers at interfaces[82] and in sonicated solutions,[102] although in the latter case the contribution of S—S bond homolysis mediated by sonolytically generated radicals[103] to the observed chemistry remains to be quantified (Fig. 10).

4.2 Emerging Trends in Contemporary Empirical Studies

The bimolecular nature of S_N2 reactions complicates both accurate quantitation of mechanochemical kinetics and its molecular interpretation. Unimolecular reactions are free of these complications and offer advantages both for fundamental studies of the effect of anisotropic strain on chemical reactivity and its exploitation. The two most extensively studied reactions are isomerizations of dihalocyclopropanes and spiropyrans (Fig. 11). Isomerization of multiple dihalocyclopropane moieties in a single chain increases its length considerably[104] and generates allylhalides that are susceptible to nucleophilic displacement. As a result, blends of poly(dichlorocyclopropanes) and polymers with carboxylic groups in side chains undergo self-strengthening in shear.[24] Mechanochemical isomerization of dihalocyclopropane has been studied extensively as an example of force-induced violation of the orbital conservation rules.[46] Spiropyrans are currently the most popular mechanochromic moiety for use in polymer, in part, due to the very low loads needed to induce color,[74] which results from thermodynamically or kinetically controlled isomerization of colorless spiropyran to colored merocyanine

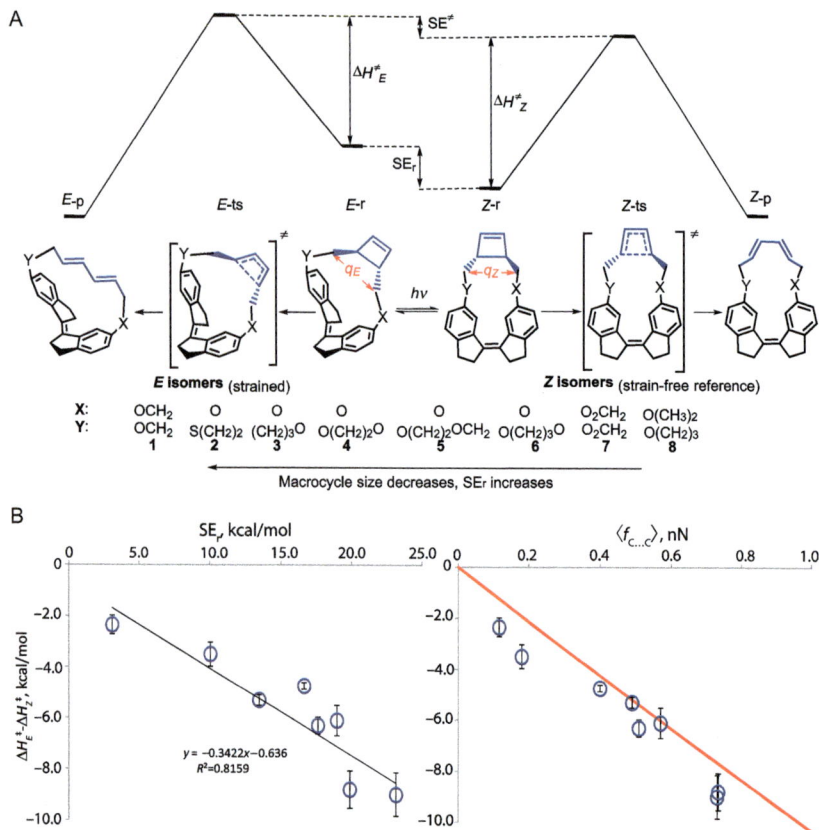

Fig. 10 Illustration of the use of stiff stilbene to measure intrinsic mechanochemical kinetics of a classical electrocyclic reaction, isomerization of *trans*-3,4-dialkylcyclobutene to a diene (*blue*) using a series of stiff stilbene macrocycles (A). The measured activation enthalpy is comparable across the series for *Z* macrocycles but decreases with decreasing macrocycle size for *E* analogs. The measured difference in activation enthalpies between the strain-free *Z*-isomer and strained *E* analog of the same macrocycle, $\Delta H^{\ddagger}_E - \Delta H^{\ddagger}_Z$, correlates well both with the calculated strain energy difference of the two isomers (B) and the restoring force of the nonbonding exocyclic C—C distance of the reactive moiety (defined by *red arrows* in (A)), (C). The *red line* in (C) is the activation enthalpy of the same reactive moiety in a stretched polymer segment with the same restoring force of the local coordinate as in an *E* macrocycle. *Reproduced with permission from Tian YC, Boulatov R. Comment on Stauch T, Dreuw A. "Stiff-stilbene photoswitch ruptures bonds not by pulling but by local heating". Phys Chem Chem Phys. 2016;18:15848. Phys Chem Chem Phys. 2016;18 (38):26990–26993.*

(other approaches to realizing mechanochromism in polymer materials are reviewed in Refs.[22,26]). Spiropyran isomerization is one of the few reactions where anisotropy of mechanochemical activation was studied, albeit qualitatively.

A

B

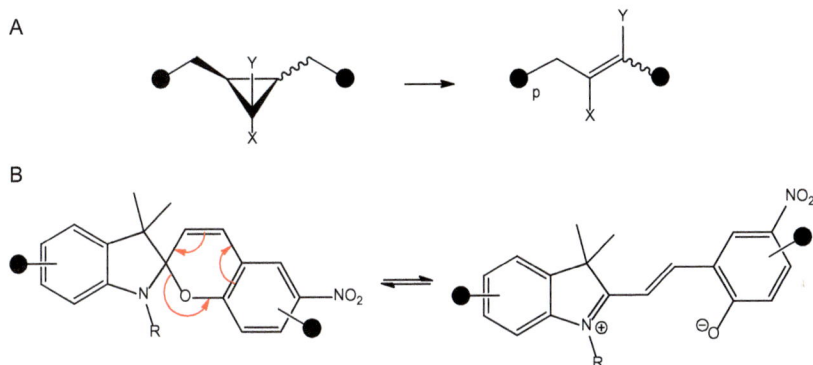

Fig. 11 The two most commonly studied mechanochemical reactions: isomerization of dihalocyclopropanes (A), where X=Y=Cl or Br or X=F, Y=Cl; and of spiropyrans (B). *Black spheres* signify point of attachment to macromolecules.

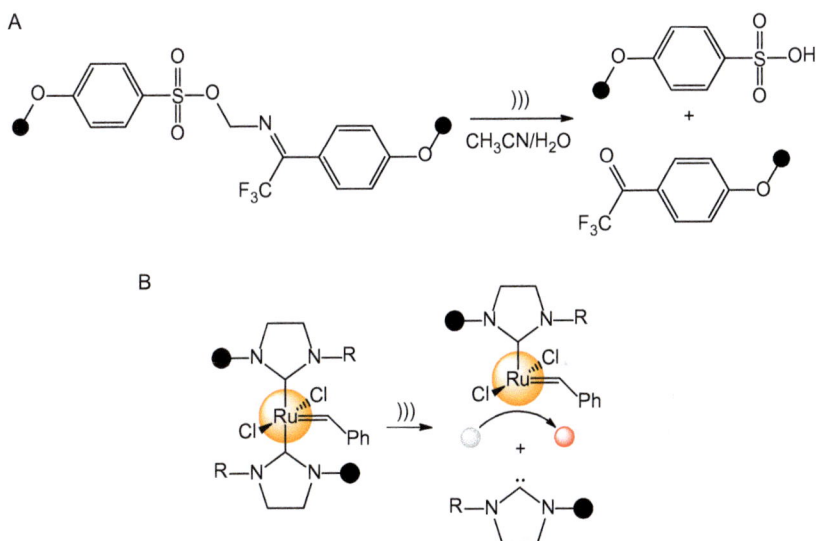

A

B

Fig. 12 Two examples of mechanochemical reaction cascades: the generation of a mechanoacid, in analogy to photoacids used in photolithography[111] and of a catalyst (for ring-closing metathesis).[106]

Design of mechanochemical reaction cascades has emerged as one of the more frequently pursued goals in contemporary polymer mechanochemistry. In such cascades, the product of a mechanochemical reaction is a reactant, initiator,[28,105–109] or catalyst[24,27,110,111] of subsequent nonmechanochemical reaction(s) (Fig. 12). An alternative to a mechanochemical cascade[63] is reaction gating whereby a reactive site blocks transmission of applied force to

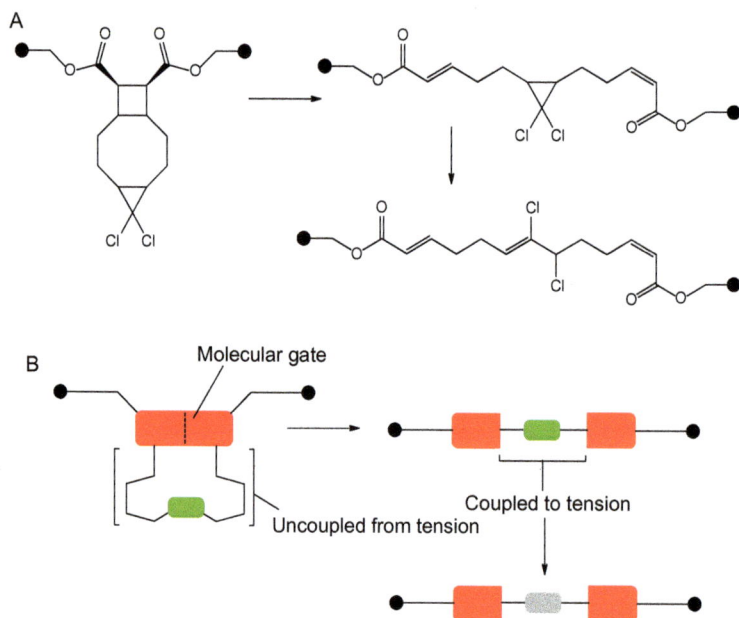

Fig. 13 An example of a single-molecule cascade or mechanochemical reaction gating. (A) The reactive site undergoes two sequential transformations: mechanochemical dissociation of the gate, which allows applied force to be transmitted to the second-reactive site (dichlorocyclopropane), which isomerizes as soon as the gate is opened. This sequence ensures that the force at which the protected reactive site reacts is determined by the intrinsic mechanochemical reactivity of the gate rather than the reaction itself. (B) A cartoon representation of the principle of mechanochemical gating.

another site until the first one reacts. In the only demonstration to date, the gate used was a cyclobutane derivative and the protected site dichlorocyclopropane (Fig. 13). Thermal dissociation of strain-free cyclobutanes to two olefins is negligibly slow, but is accelerated by tensile loading, as is isomerization of dihalocyclopropanes. Cyclobutane, however, withstands much higher tensile load than dichrolocyclopropane, so that the threshold force at which the latter isomerizes is determined by the mechanochemical kinetics of the gating reaction instead of the substrate (dichlorocyclopropane). Another conceptually interesting and promising direction has been integration of multiple productive responses to anisotropic strain in a single reactive moiety, resulting in so-called multimodal "mechanophores."[25,27]

5. SUMMARY

Polymer mechanochemistry is an emerging discipline at the interface of chemistry, physics, and engineering, which aims at understanding and exploiting unique reactivity that becomes accessible when a polymer chain is overstretched.[1] Polymer chains become overstretched in a variety of technologically important processes, and the reactivity of such overstretched chains often determines the bulk response of the material to mechanical load, including catastrophic material failure and more gradual (but no less detrimental) material aging. In the laboratory, micromanipulation techniques and flow fields allow individual macromolecular chains to be stretched with some degree of control over the magnitude and the duration of the imposed strain, and the rate at which it is imposed.

The chemical consequences of stretching polymer chains are typically discussed in terms of the effect of force on reactivity. The reason is that macromolecular chains are most often stretched when a macroscopic (or microscopic) object moves directionally, be it arm movement that stretches a rubber band, the retraction of an AFM tip that stretches a single macromolecule connecting it to another surface, or the rapid flow of a polymer solution through narrow channels. Force is a variable that allows, at least in theory, quantitative description of both macroscopic motion and the effect of this motion on reaction dynamics.[11,52] For example, force allows us to extrapolate the kinetic stability of a stretched chain of an arbitrary length from the reactivity of a single monomer, either calculated or observed in properly designed macrocycles.[25,58,63]

The outcome of overstretching a simple polymer such as polystyrene or polyacrylate is rather boring, as the chain simply fragments by homolysis of a backbone bond (although the resulting macroradicals may manifest rich chemistry).[1] Modern synthetic methods allow diverse reactive moieties to be decorated with two or more macromolecular chains so that when the resulting polymer is stretched, large and anisotropic strain is imposed on the reactive site. Likewise, multiple reactive sites can be connected in series by mechanochemically "inert" linkers.[32] Stretching such polymers yields diverse chemistry, from site-selective fragmentation, to chemomechanoluminescence[21] and stabilization of structures that are transition states in strain-free reactive sites.[112] Evidence suggests that some reactions proceed by mechanisms that are kinetically negligible or through intermediates or transition states that do not exist in the absence of strain.[42,46] Attaching polymer chains to different pairs of atoms of the same reactive moiety enables one to study anisotropy of strain/reactivity relationship.[6] Mechanochemical reactions have been used to control other reactions, either by producing well-defined reactive species[26,27,110] or redistributing imposed strain.[63]

Although it may seem intuitive that stretching a molecule would accelerate its fragmentation along the stretching axis, the kinetic response of reactive sites to such anisotropic straining is far richer than that. Both computations and experiment indicate that stretching a molecule along one axis either inhibits its fragmentation along this axis, or accelerates its fragmentation along an orthogonal axis.[42] Such responses lack functional analogies in the macroscopic world and do not seem to follow the FERs of physical organic chemistry[51] but are amenable to quantitative predictions within the formalism of local restoring force.

This proliferation of empirical mechanochemical data offers physical organic chemists a great opportunity to influence the evolution of the field by helping to discover the mechanisms of mechanochemical reactions, which remain little studied, and exploiting this knowledge to guide the design of new mechanochemical reactions. Likewise, while we can calculate the force-dependent activation barriers of many reactions, we do not yet know how accurate the results are in general: for certain reactions quantum chemical calculations reproduce experimental measurements quantitatively[25,58,63]; whereas for others they are even qualitatively incorrect. We need to learn how to extract better quality quantitative molecular data from experimental techniques of polymer mechanochemistry, particularly sonication, and expand the range of molecular architectures that reproduce the highly anisotropic strains imposed on small reactive sites in stretched polymers without the complexity of coupled macroscopic motion.

Studying overstretched polymer chains offers an opportunity to greatly expand our understanding of chemical reactivity, particularly of highly strained molecular geometries not accessible synthetically, and to create new materials with unique modes of response to mechanical loads.[23] Physical organic chemistry has much offer to realize this opportunity.

REFERENCES

1. Akbulatov S, Boulatov R. Experimental polymer mechanochemistry and its interpretational frameworks. *ChemPhysChem*. 2017;18(11):1422–1450.
2. Kucharski TJ, Tian YC, Akbulatov S, Boulatov R. Chemical solutions for the closed-cycle storage of solar energy. *Energy Environ Sci*. 2011;4(11):4449–4472.
3. Liebman JF, Greenberg A. Survey of strained organic-molecules. *Chem Rev*. 1976;76(3):311–365.
4. Katz TJ, Acton N. Synthesis of prismane. *J Am Chem Soc*. 1973;95(8):2738–2739.
5. Robb MJ, Kim TA, Halmes AJ, White SR, Sottos NR, Moore JS. Regioisomer-specific mechanochromism of naphthopyran in polymeric materials. *J Am Chem Soc*. 2016;138(38):12328–12331.
6. Stevenson R, De Bo G. Controlling reactivity by geometry in retro–Diels–Alder reactions under tension. *J Am Chem Soc*. 2017;139(46):16768–16771.
7. Cramer CJ. *Essentials of Computational Chemistry*. 2nd ed. Chichester: Wiley; 2004.

8. Klippenstein SJ, Pande VS, Truhlar DG. Chemical kinetics and mechanisms of complex systems: a perspective on recent theoretical advances. *J Am Chem Soc.* 2014;136:528–546.

9. Cintas P, Cravotto G, Barge A, Marina K. Interplay between mechanochemistry and sonochemistry. In: Boulatov R, ed. *Polymer Mechanochemistry.* Cham: Springer International Publishing; 2015:239–284.

10. Chen ZX, Mercer JAM, Zhu XL, et al. Mechanochemical unzipping of insulating polyladderene to semiconducting polyacetylene. *Science.* 2017;357(6350):475–478.

11. Boulatov R. Reaction dynamics in the formidable gap. *Pure Appl Chem.* 2011;83(1): 25–41.

12. Huang Z, Boulatov R. Chemomechanics: chemical kinetics for multiscale phenomena. *Chem Soc Rev.* 2011;40(5):2359–2384.

13. Berry RS, Rice SA, Ross J. *Physical and Chemical Kinetics.* 2nd ed. vol 3. New York: Oxford University Press; 2002.

14. Fleming GR, Ratner MA. *Directing Matter and Energy: Five Challenges for Science and the Imagination: A Report From the Basic Energy Sciences Advisory Committee.* U.S. Department of Energy, Office of Basic Energy Sciences; 2007.

15. Zhang H, Lin YJ, Xu YZ, Weng WG. Mechanochemistry of topological complex polymer systems. In: Boulatov R, ed. *Polymer Mechanochemistry.* vol 369. 2015:135–207.

16. Zhang QM, Serpe MJ. Responsive polymers as sensors, muscles and self-healing materials. In: Boulatov R, ed. *Polymer Mechanochemistry.* Cham: Springer International Publishing; 2015:377–424.

17. Patrick JF, Robb MJ, Sottos NR, Moore JS, White SR. Polymers with autonomous life-cycle control. *Nature.* 2016;540(7633):363–370.

18. Gostl R, Sijbesma RP. Pi-extended anthracenes as sensitive probes for mechanical stress. *Chem Sci.* 2016;7(1):370–375.

19. Gossweiler GR, Hewage GB, Soriano G, et al. Mechanochemical activation of covalent bonds in polymers with full and repeatable macroscopic shape recovery. *ACS Macro Lett.* 2014;3(3):216–219.

20. Zhang H, Chen YJ, Lin YJ, et al. Spiropyran as a mechanochromic probe in dual cross-linked elastomers. *Macromolecules.* 2014;47(19):6783–6790.

21. Ducrot E, Chen Y, Bulters M, Sijbesma RP, Creton C. Toughening elastomers with sacrificial bonds and watching them break. *Science.* 2014;344(6180):186–189.

22. Haehnel AP, Sagara Y, Simon YC, Weder C. Mechanochemistry in polymers with supramolecular mechanophores. In: Boulatov R, ed. *Polymer Mechanochemistry.* Cham: Springer International Publishing; 2015:345–375.

23. Boulatov R. The challenges and opportunities of contemporary polymer mechano-chemistry. *ChemPhysChem.* 2017;18(11):1419–1421.

24. Ramirez ALB, Kean ZS, Orlicki JA, et al. Mechanochemical strengthening of a synthetic polymer in response to typically destructive shear forces. *Nat Chem.* 2013;5(9):757–761.

25. Zhang H, Li X, Lin YJ, et al. Multi-modal mechanophores based on cinnamate dimers. *Nat Commun.* 2017;8:1147.

26. Clough JM, Balan A, Sijbesma RP. Mechanochemical reactions reporting and repairing bond scission in polymers. In: Boulatov R, ed. *Polymer Mechanochemistry.* Cham: Springer International Publishing; 2015:209–238.

27. Zhang H, Gao F, Cao XD, et al. Mechanochromism and mechanical-force-triggered cross-linking from a single reactive moiety incorporated into polymer chains. *Angew Chem Int Ed.* 2016;55(9):3040–3044.

28. Imato K, Otsuka H. Reorganizable and stimuli-responsive polymers based on dynamic carbon-carbon linkages in diarylbibenzofuranones. *Polymer.* 2018;137:395–413.

29. Hirokawa N. Kinesin and dynein superfamily proteins and the mechanism of organelle transport. *Science.* 1998;279(5350):519–526.

30. Marszalek PE, Lu H, Li HB, et al. Mechanical unfolding intermediates in titin modules. *Nature*. 1999;402(6757):100–103.

31. Astumian RD. Huxley's model for muscle contraction revisited: the importance of microscopic reversibility. In: Boulatov R, ed. *Polymer Mechanochemistry*. Springer International Publishing; 2015:285–316. 369.

32. Bowser BH, Craig SL. Empowering mechanochemistry with multi-mechanophore polymer architectures. *Polym Chem*. 2018;9:3583–3593. advanced article.

33. Sotomayor M, Schulten K. Single-molecule experiments in vitro and in silico. *Science*. 2007;316(5828):1144.

34. De Vico L, Liu YJ, Krogh JW, Lindh R. Chemiluminescence of 1,2-dioxetane. Reaction mechanism uncovered. *J Phys Chem A*. 2007;111(32):8013–8019.

35. Farahani P, Roca-Sanjuan D, Zapata F, Lindh R. Revisiting the nonadiabatic process in 1,2-dioxetane. *J Chem Theory Comput*. 2013;9(12):5404–5411.

36. Kauzmann W, Eyring H. The viscous flow of large molecules. *J Am Chem Soc*. 1940;62:3113–3125.

37. Bell GI. Models for specific adhesion of cells to cells. *Science*. 1978;200(4342):618–627.

38. Evans E, Ritchie K. Dynamic strength of molecular adhesion bonds. *Biophys J*. 1997;72(4):1541–1555.

39. Kramers HA. Brownian motion in a field of force and the diffusion model of chemical reactions. *Phys Ther*. 1940;7:284–304.

40. Hanggi P, Talkner P, Borkovec M. Reaction-rate theory—50 years after kramers. *Rev Mod Phys*. 1990;62(2):251–341.

41. Tian YC, Boulatov R. Quantum-chemical validation of the local assumption of chemomechanics for a unimolecular reaction. *ChemPhysChem*. 2012;13(9):2277–2281.

42. Akbulatov S, Tian Y, Huang Z, Kucharski TJ, Yang QZ, Boulatov R. Experimentally realized mechanochemistry distinct from force-accelerated scission of loaded bonds. *Science*. 2017;357(6348):299–303.

43. Kochhar GS, Heverly-Coulson GS, Mosey NJ. Theoretical approaches for understanding the interplay between stress and chemical reactivity. In: Boulatov R, ed. *Polymer Mechanochemistry*. vol 369. Cham: Springer International Publishing; 2015:37–96.

44. Ribas-Arino J, Marx D. Covalent mechanochemistry: theoretical concepts and computational tools with applications to molecular nanomechanics. *Chem Rev*. 2012;112(10):5412–5487.

45. Hammond GS. A correlation of reaction rates. *J Am Chem Soc*. 1955;77(2):334–338.

46. Wang JP, Kouznetsova TB, Niu ZB, et al. Inducing and quantifying forbidden reactivity with single-molecule polymer mechanochemistry. *Nat Chem*. 2015;7(4):323–327.

47. Marcus RA, Sutin N. Electron transfers in chemistry and biology. *Biochim Biophys Acta*. 1985;811(3):265–322.

48. Marcus RA. Electron-transfer reactions in chemistry—theory and experiment. *Rev Mod Phys*. 1993;65(3):599–610.

49. Brauman JI, Dodd JA, Han C-C. Intrinsic nucleophilicity. In: Harris J, ed. *Nucleophilicity*. Washington, DC: American Chemical Society; 1987.

50. Dudko OK. Decoding the mechanical fingerprints of biomolecules. *Q Rev Biophys*. 2016;49(3), e3.

51. Jencks WP. A primer for the Bema Hapothle—an empirical-approach to the characterization of changing transition-state structures. *Chem Rev*. 1985;85(6):511–527.

52. Kucharski TJ, Boulatov R. The physical chemistry of mechanoresponsive polymers. *J Mater Chem*. 2011;21(23):8237–8255.

53. Huang Z, Yang QZ, Khvostichenko D, Kucharski TJ, Chen J, Boulatov R. Method to derive restoring forces of strained molecules from kinetic measurements. *J Am Chem Soc*. 2009;131(4):1407–1409.

54. Kucharski TJ, Huang Z, Yang QZ, et al. Kinetics of thiol/disulfide exchange correlate weakly with the restoring force in the disulfide moiety. *Angew Chem Int Ed.* 2009;48(38):7040–7043.

55. Yang QZ, Huang Z, Kucharski TJ, Khvostichenko D, Chen J, Boulatov R. A molecular force probe. *Nat Nanotechnol.* 2009;4(5):302–306.

56. Kucharski TJ, Yang QZ, Tian YC, Boulatov R. Strain-dependent acceleration of a paradigmatic S(N)2 reaction accurately predicted by the force formalism. *J Phys Chem Lett.* 2010;1(19):2820–2825.

57. Hermes M, Boulatov R. The entropic and enthalpic contributions to force-dependent dissociation kinetics of the pyrophosphate bond. *J Am Chem Soc.* 2011;133(50):20044–20047.

58. Akbulatov S, Tian YC, Boulatov R. Force-reactivity property of a single monomer is sufficient to predict the micromechanical behavior of its polymer. *J Am Chem Soc.* 2012;134(18):7620–7623.

59. Akbulatov S, Tian YC, Kapustin E, Boulatov R. Model studies of the kinetics of ester hydrolysis under stretching force. *Angew Chem Int Ed.* 2013;52(27):6992–6995.

60. Tian YC, Boulatov R. Comparison of the predictive performance of the Bell–Evans, Taylor-expansion and statistical-mechanics models of mechanochemistry. *Chem Commun.* 2013;49(39):4187–4189.

61. Tian YC, Kucharski TJ, Yang QZ, Boulatov R. Model studies of force-dependent kinetics of multi-barrier reactions. *Nat Commun.* 2013;4:2538.

62. Huang Z, Boulatov R. Chemomechanics with molecular force probes. *Pure Appl Chem* 2010;82(4):931–951.

63. Wang JP, Kouznetsova TB, Boulatov R, Craig SL. Mechanical gating of a mechanochemical reaction cascade. *Nat Commun.* 2016;7:13433.

64. Boulatov R. Demonstrated leverage. *Nat Chem.* 2013;5(2):84–86.

65. Klukovich HM, Kouznetsova TB, Kean ZS, Lenhardt JM, Craig SL. A backbone lever-arm effect enhances polymer mechanochemistry. *Nat Chem.* 2013;5(2):110–114.

66. Dopieralski P, Ribas-Arino J, Anjukandi P, Krupicka M, Marx D. Force-induced reversal of beta-eliminations: stressed disulfide bonds in alkaline solution. *Angew Chem Int Ed.* 2016;55(4):1304–1308.

67. Wang JP, Kouznetsova TB, Craig SL. Reactivity and mechanism of a mechanically activated anti-Woodward–Hoffmann–DePuy reaction. *J Am Chem Soc.* 2015;137(36):11554–11557.

68. Cheng B, Cui SX. Supramolecular chemistry and mechanochemistry of macromolecules: recent advances by single-molecule force spectroscopy. In: Boulatov R, ed. *Polymer Mechanochemistry.* vol 369. Springer; 2015:97–134.

69. Kouznetsova TB, Wang JP, Craig SL. Combined constant-force and constant-velocity single-molecule force spectroscopy of the conrotatory ring opening reaction of benzocyclobutene. *ChemPhysChem.* 2017;18(11):1486–1489.

70. Kersey FR, Yount WC, Craig SL. Single-molecule force spectroscopy of bimolecular reactions: system homology in the mechanical activation of ligand substitution reactions. *J Am Chem Soc.* 2006;128(12):3886–3887.

71. Schmidt SW, Beyer MK, Clausen-Schaumann H. Dynamic strength of the silicon–carbon bond observed over three decades of force-loading rates. *J Am Chem Soc.* 2008;130:3664.

72. Hummer G. Nonequilibrium methods for equilibrium free energy calculations. In: Chipot C, Pohorille A, eds. *Free Energy Calculations.* Berlin: Springer; 2007.

73. Minkin VI. Photo-, thermo-, solvato-, and electrochromic spiroheterocyclic compounds. *Chem Rev.* 2004;104(5):2751–2776.

74. Gossweiler GR, Kouznetsova TB, Craig SL. Force-rate characterization of two spiropyran-based molecular force probes. *J Am Chem Soc.* 2015;137(19):6148–6151.

75. Moffitt JR, Chemla YR, Smith SB, Bustamante C. Recent advances in optical tweezers. *Annu Rev Biochem.* 2008;77:205–228.
76. Tirtaatmadja V, Sridhar T. A filament stretching device for measurement of extensional viscosity. *J Rheol.* 1993;37(6):1081–1102.
77. Taylor GI. The formation of emulsions in definable fields of flow. *Proc R Soc London Ser A.* 1934;146(A858):0501–0523. Containing Papers of a Mathematical and Physical Character.
78. Hsieh C-C, Park SJ, Larson RG. Brownian dynamics modeling of flow-induced birefringence and chain scission in dilute polymer solutions in a planar cross-slot flow. *Macromolecules.* 2005;38(4):1456–1468.
79. Larson RG. The rheology of dilute solutions of flexible polymers: progress and problems. *J Rheol.* 2005;49(1):1–70.
80. Perkins TT, Smith DE, Chu S. Single polymer dynamics in an elongational flow. *Science.* 1997;276(5321):2016–2021.
81. Scrivener O, Berner C, Cressely R, Hocquart R, Sellin R, Vlachos NS. Dynamical behavior of drag-reducing polymer-solutions. *J Non-Newtonian Fluid Mech.* 1979;5(APR):475–495.
82. Li YC, Sheiko SS. Molecular mechanochemistry: engineering and implications of inherently strained architectures. In: Boulatov R, ed. *Polymer Mechanochemistry.* vol 369. Springer; 2015:1–36.
83. Yan X, Xu J-F, Cook TR, et al. Photoinduced transformations of stiff-stilbene-based discrete metallacycles to metallosupramolecular polymers. *Proc Natl Acad Sci.* 2014;111(24):8717.
84. Wang Y, Tian Y, Chen Y-Z, et al. A light-driven molecular machine based on stiff stilbene. *Chem Commun.* 2018;54:7991–7994.
85. Kean ZS, Akbulatov S, Tian YC, Widenhoefer RA, Boulatov R, Craig SL. Photomechanical actuation of ligand geometry in enantioselective catalysis. *Angew Chem Int Ed.* 2014;53(52):14508–14511.
86. Kucharski TJ, Boulatov R. Fundamentals of molecular photoactuation. In: Knopf GK, Otani Y, eds. *Optical Nano and Micro Actuator Technology.* CRC Press; 2012:83–106. ISBN: 978-1-4398-4054-2. [Chapter 3].
87. Tian YC, Boulatov R. Comment on Stauch T, Dreuw A. "Stiff-stilbene photoswitch ruptures bonds not by pulling but by local heating". Phys Chem Chem Phys. 2016;18:15848. *Phys Chem Chem Phys.* 2016;18(38):26990–26993.
88. Carpenter BK. Nonstatistical dynamics in thermal reactions of polyatomic molecules. *Annu Rev Phys Chem.* 2004;56(1):57–89.
89. Kean ZS, Niu Z, Hewage GB, Rheingold AL, Craig SL. Stress-responsive polymers containing cyclobutane core mechanophores: reactivity and mechanistic insights. *J Am Chem Soc.* 2013;135(36):13598–13604.
90. Wiita AP, Ainavarapu RK, Huang HH, Fernandez JM. Force-dependent chemical kinetics of disulfide bond reduction observed with single-molecule techniques. *Proc Natl Acad Sci U S A.* 2006;103(19):7222–7227.
91. Liang J, Fernandez JM. Mechanochemistry: one bond at a time. *ACS Nano.* 2009;3(7):1628–1645.
92. Singh R, Whitesides GM. Reagents for rapid reduction of native disulfide bonds in proteins. *Bioorg Chem.* 1994;22(1):109–115.
93. Singh R, Whitesides GM. Thiol-disulfide interchange. In: Patai S, Rappoport Z, eds. *The Chemistry of Sulphur-Containing Functional Groups, Supplement S.* 1993:633–658. ISBN: 978-0-470-03440-8. [Chapter 13].
94. Li WJ, Grater F. Atomistic evidence of how force dynamically regulates thiol/disulfide exchange. *J Am Chem Soc.* 2010;132(47):16790–16795.

95. Iozzi MF, Helgaker T, Uggerud E. Influence of external force on properties and reactivity of disulfide bonds. *J Phys Chem A*. 2011;115(11):2308–2315.

96. Baldus IB, Grater F. Mechanical force can fine-tune redox potentials of disulfide bonds. *Biophys J*. 2012;102(3):622–629.

97. Hofbauer F, Frank I. CPMD simulation of a bimolecular chemical reaction: nucleophilic attack of a disulfide bond under mechanical stress. *Chem Eur J*. 2012;18(51): 16332–16338.

98. Dopieralski P, Ribas-Arino J, Anjukandi P, Krupicka M, Kiss J, Marx D. The Janus-faced role of external forces in mechanochemical disulfide bond cleavage. *Nat Chem*. 2013;5(8):685–691.

99. Peverati R, Truhlar DG. Quest for a universal density functional: the accuracy of density functionals across a broad spectrum of databases in chemistry and physics. *Philos Trans R Soc A Math Phys Eng Sci*. 2014;372(2011):20120476.

100. Cheong PH-Y, Legault CY, Um JM, Çelebi-Ölçüm N, Houk KN. Quantum mechanical investigations of organocatalysis: mechanisms, reactivities, and selectivities. *Chem Rev*. 2011;111(8):5042–5137.

101. Chen M, Zheng L, Santra B, et al. Hydroxide diffuses slower than hydronium in water because its solvated structure inhibits correlated proton transfer. *Nat Chem*. 2018;10(4):413–419.

102. Fritze UF, Craig SL, von Delius M. Disulfide-centered poly(methyl acrylates): four different stimuli to cleave a polymer. *J Polym Sci A Polym Chem*. 2018;56(13): 1404–1411.

103. Fritze UF, von Delius M. Dynamic disulfide metathesis induced by ultrasound. *Chem Commun*. 2016;52(38):6363–6366.

104. Wu D, Lenhardt JM, Black AL, Akhremitchev BB, Craig SL. Molecular stress relief through a force-induced irreversible extension in polymer contour length. *J Am Chem Soc*. 2010;132(45):15936–15938.

105. Piermattei A, Karthikeyan S, Sijbesma RP. Activating catalysts with mechanical force. *Nat Chem*. 2009;1(2):133–137.

106. Jakobs RTM, Sijbesma RP. Mechanical activation of a latent olefin metathesis catalyst and persistence of its active species in ROMP. *Organometallics*. 2012;31(6):2476–2481.

107. Michael P, Binder WH. A Mechanochemically triggered "click" catalyst. *Angew Chem Int Ed*. 2015;54(47):13918–13922.

108. Wang JP, Piskun I, Craig SL. Mechanochemical strengthening of a multi-mechanophore benzocyclobutene polymer. *ACS Macro Lett*. 2015;4(8):834–837.

109. Clough JM, Balan A, van Daal TJ, Sijbesma RP. Probing force with mechanobase-induced chemiluminescence. *Angew Chem Int Ed*. 2016;55(4):1445–1449.

110. Verstraeten F, Gostl R, Sijbesma RP. Stress-induced colouration and crosslinking of polymeric materials by mechanochemical formation of triphenylimidazolyl radicals. *Chem Commun*. 2016;52(55):8608–8611.

111. Nagamani C, Liu HY, Moore JS. Mechanogeneration of acid from oxime sulfonates. *J Am Chem Soc*. 2016;138(8):2540–2543.

112. Lenhardt JM, Ong MT, Choe R, Evenhuis CR, Martinez TJ, Craig SL. Trapping a diradical transition state by mechanochemical polymer extension. *Science*. 2010;329(5995): 1057–1060.

CPI Antony Rowe
Chippenham, UK
2019-01-02 16:59